FROM HYSTERIA TO HORMONES

RSA·STR

THE **RSA** SERIES IN TRANSDISCIPLINARY **RHETORIC**

The RSA Series in Transdisciplinary Rhetoric is a collaboration with the Rhetoric Society of America to publish innovative and rigorously argued scholarship on the tremendous disciplinary breadth of rhetoric. Books in the series take a variety of approaches, including theoretical, historical, interpretive, critical, or ethnographic, and examine rhetorical action in a way that appeals, first, to scholars in communication studies and English or writing, and, second, to at least one other discipline or subject area.

Amy Koerber

FROM HYSTERIA TO HORMONES

A Rhetorical History

THE PENNSYLVANIA STATE UNIVERSITY PRESS
UNIVERSITY PARK, PENNSYLVANIA

Library of Congress Cataloging-in-Publication Data

Names: Koerber, Amy (Amy Lunn), author.
Title: From hysteria to hormones : a rhetorical history / Amy
    Koerber.
Other titles: RSA series in transdisciplinary rhetoric.
Description: University Park, Pennsylvania : The Pennsylvania
    State University Press, [2018] | Series: The RSA series in
    transdisciplinary rhetoric | Includes bibliographical
    references and index.
Summary: "Examines the rhetorical activity that preceded the
    early twentieth-century emergence of the word 'hormone'
    and the impact of this word on expert understandings of
    women's health"—Provided by publisher.
Identifiers: LCCN 2017050844 | ISBN 9780271080857
    (cloth : alk. paper)
Subjects: LCSH: Women—Health and hygiene—History. |
    Hormones—History. | Hysteria—History. | Rhetoric.
Classification: LCC RA564.85 .K655 2018 | DDC
    613/.04244—dc23
LC record available at https://lccn.loc.gov/2017050844

The Pennsylvania State University Press is a member of the
Association of University Presses.

It is the policy of The Pennsylvania State University Press to
use acid-free paper. Publications on uncoated stock satisfy the
minimum requirements of the American National Standard
for Information Sciences—Permanence of Paper for Printed
Library Material, ANSI Z39.48–1992.

To my sisters:

Anna, Marne, Kristen

Contents

Illustrations

Preface

But for Adam no suitable helper was found. So the Lord God caused the man to fall into a deep sleep; and while he was sleeping, he took one of the man's ribs and then closed up the place with flesh.

Then the Lord God made a woman from the rib he had taken out of the man, and he brought her to the man.

The man said, "this is now bone of my bones and flesh of my flesh; she shall be called 'woman,' for she was taken out of man."

That is why a man leaves his father and mother and is united to his wife, and they become one flesh.

Adam and his wife were both naked, and they felt no shame.

—Genesis 2:21–25

After God so graciously created a helper for Adam, the man and woman lived together in pure bliss for a little while. Then one day, so the story goes, "when the woman saw that the fruit of the tree was good for food and pleasing to the eye, and also desirable for gaining wisdom, she took some and ate it. She also gave some to her husband, who was with her, and he ate it."[1] And we all know the rest of that story.

Whether through a religious tale about the first woman who could not resist a juicy piece of fruit, or through using the most sophisticated scientific techniques available in the twenty-first century, the effort to find language that accounts for the differences between men and women has been relentless, in the Western world and elsewhere. This quest has taken us down many different roads, some of which have turned out to be dead ends, and many of which seem pretty bizarre from today's perspective.

For example, the predominant belief about sex difference in ancient Greece was that the reproductive organs, both male and female, were wild animals. However, there was an important difference between these animals' behaviors in male and female bodies. The man's "wild animal" existed in a fixed and visible location outside his body. The woman's, by contrast, lived inside her body, where it was free to roam around unseen and cause unexpected problems. Historian

Nancy Demand explains that according to one of the popular medical theories at the time, women were believed to be "weaker, moister, softer, more porous, and warmer than men," and females were thought to be incomplete versions of males. Furthermore, it was believed that "without the moisture and weight provided by semen and the fetus, the womb would wander about the body causing alarming and dangerous symptoms."[2] In other words, women in ancient Greece who were not pregnant were seen as subject to all manner of health problems caused by the dangerous, yet unseen, wandering womb. Recommended treatments for the wandering womb included "fumigations and odor therapies"[3] that would attract the animalistic womb back to its natural location. Intercourse was also frequently prescribed as a cure for such problems as menstrual blockages.

The first use of the term *hysteria* to describe this condition is usually attributed to Hippocrates, who used the word "hysteron" (literally, "movement of the uterus," but derived from the Sanskrit word for *belly*) to explain the origins of a number of women's health symptoms.[4] Concerns about the numerous health problems created by the wandering womb persisted for many centuries, although as early as the thirteenth century, Western physicians started to question whether hysteria was literally caused by movement of the uterus, or whether the female brain could also possibly have been involved. Within the confines of science and technology available for studying the human body at that time, of course, such questioning of these ancient Greek beliefs could only exist in the realm of speculation, and ancient Greek ideas about the uterus as a physical cause for hysteria persisted quite explicitly in many of the medical texts and widely accepted customs of the nineteenth and early twentieth centuries. One well-known example is Victorian women's practice of carrying smelling salts, a practice which recalled the idea that a pleasurable odor could restore the wandering uterus to its natural place.[5]

In the twenty-first century, of course, we have achieved a more scientific understanding of the human female's anatomy. This does not mean that science has finally discovered the truth about male-female difference, but rather that we have mechanisms such as double-blind peer review and competition for grant funding to ensure that if someone claims that women's reproductive organs are wild animals wandering uncontrollably inside their bodies, then those claims will not go unchallenged, and such claims will not be convincing to anyone if there is no evidence to support them. As a result, today it is unlikely that a physician will attribute a female patient's symptoms to the fact that her uterus is a wild animal that refuses to stay in place. We have ultrasounds to visualize the internal anatomy of male and female bodies and to diagnose underlying problems in a much more scientifically verifiable way than has ever been possible before in the history of medicine. We have blood tests to determine levels of various bodily substances that might explain some health conditions. And we have images produced by brain scans to provide evidence of the differences between men's and women's brains. We even have entirely disparate medical specialties devoted to the disorders that might afflict the brain and the uterus, and physicians in each of these specialties have access to their own array of pharmaceutical products that can be used to restore order from the outside when the body's own internal systems (whether neurological, hormonal, or otherwise) seem to be malfunctioning.

The early twentieth-century discovery of hormones certainly has an important role to play in this transformation from mythical to scientific understandings of women's bodies and the health problems that they experience. Shortly after Ernest Henry Starling coined the term *hormone* in 1905, hormones began to provide a chemical explanation for bodily phenomena that were, for many centuries prior, understood in terms of vague, unscientific notions such as wandering wombs, humors, energies, and balance.[6] By the 1930s, endocrinology was established as a discipline, and experts could offer scientific explanations that referred to specific reproductive hormones as the causes for female symptoms that had for centuries been vaguely defined and often lumped together under the unspecific, constantly changing diagnosis of hysteria.[7] These hormones were first identified vaguely under the umbrella term "female sex hormones," but they later came to be known more specifically as estrogen and progesterone. In short, hormones are the scientific entity most prominent in replacing hysteria as a catch-all diagnosis for female problems.

In contrast to previous scholars' emphasis on the profound changes that occurred after the discovery of hormones,[8] I argue in this book that the discovery of hormones was not so much a revolution as an exigency that required old ways of thinking about women's bodies to be twisted, reshaped, and transformed to fit the new turn-of-century expectations that medical practices and recommendations would be based in science. This revised approach to the rhetorical history of hormones takes seriously philosopher Michel Serres's admonition to "beware of philosophies that put he who practices them in the august position of always being right, of always being the wisest, the most intelligent, and the strongest."[9] Modern science, Serres reminds us, is based in just such an epistemology. We talk in terms of scientific revolution, assuming that the newest ideas represent a sharp break from the older ideas and that the newest ideas are the best. Based on many of the previous historical analyses that are available, we might be tempted to see the discovery of hormones in the early twentieth century as just such a revolutionary departure from past ways of understanding women's health. In contrast, my main argument in this book is that hormonal explanations did not necessarily replace older notions like hysteria, at least not at a clearly discernible moment in history. Rather, the rhetorical analysis presented in this book reveals that the boundary between older, nonscientific ways of understanding women's bodies and newer, scientific understandings is much murkier than we might expect.

The persistence of hysterical neurosis as a diagnostic category in the *Diagnostic and Statistical Manual of Mental Disorders* until 1994—when this term was removed from the publication's fourth edition—is just one example of the murkiness of the boundary between older and newer understandings of women's health. Furthermore, even though hysteria (and hysterical neurosis) has now been eliminated as a valid medical diagnosis, many symptoms affiliated with it remain in the current edition of the DSM.[10] Even today, much of the scientific research about women and their hormones is not as progressive as one might expect. Hormones seem to have allowed scientists to move away from theories of the womb as the main motivator of women's behaviors, but researchers have not abandoned the basic presumption that appears in ancient texts— namely, the idea that women are motivated by something inside themselves that they cannot control, whereas men control themselves through rationality and the male brain.

This glimpse at a complicated history helps explain why even in the most recent scientific literature, we still see studies that examine the uterus-brain relationship and how it might impact women's behaviors and abilities. Such studies inquire, for instance, about how the volume of regional gray matter in the brain fluctuates throughout the menstrual cycle,[11] how the "neuroplasticity" of female rodents' brains increases after they bear offspring,[12] or how women respond differently to challenging math problems depending on the phase of their menstrual cycles.[13] As these examples suggest, even in the twenty-first century, scientists in various disciplines remain interested in the ways in which the female body creates complications for the female brain, and they seem to be finding endless new ways to articulate the research questions and implement the studies that will help them better understand these effects.

A quick Google search for "hormonal woman" turns up dozens of sites that use negative language to characterize the effects of hormones on women's bodies in popular discourse. WebMD, for example, informs us that "hormonal ups and downs can wreak havoc on a woman's life," but the instructions that follow this comment in the article promise to help women "escape the horror hormones cause."[14] Another site proclaims to inform readers about "10 Things Men Should Know about Female Hormones." Hormones are described at this site as "the reason why she snaps and bites people's heads off for no apparent reason, but also why she cries while watching that sappy movie that she's already seen dozens of times."[15] And, as suggested in figure 1, the stock photos that result from an Internet search for "hormonal woman" are far from flattering.

1 | Stock image resulting from Internet image search for "hormonal woman."

Thus, I argue in this book that the gradual transformation that has occurred between the early twentieth century and the present—from hysteria to hormones—can be best understood as a rearrangement of the dominant relationship among the symptoms, causes, and diagnostic categories that we use to understand the age-old phenomenon of "female problems." Hysteria was a catchall phrase that for many centuries was used as the diagnosis for a large, diverse array of symptoms that were attributed to a wide range of causes—everything from physical symptoms like indigestion to nervous symptoms like anxiety and tension. Depending on the historical era, these symptoms had several possible etiological explanations, including the wandering womb, problems with the ovaries, a womb that is physically displaced, or something in the brain and central nervous system. When the shift to a hormonal understanding of women's health occurred over the first few decades of the twentieth century, a shift also occurred in the rhetorical mechanisms of diagnosis. Instead of one disease (hysteria) with many different symptoms and possible causes, we see a shift toward a situation in which there are many different diseases and symptoms but only one possible cause: hormones. The symptoms are the constant in this equation. They remained remarkably similar throughout the eras—always hard to pin down, and always quite diverse. These are the symptoms that physicians throughout the eras persistently refer to as "female problems" and, quite frequently, they describe these problems with terms such as "obscure" or "mysterious." As for the disease itself, what was consistently known as hysteria for several centuries started to fragment in the early twentieth century into a number of discrete diagnoses. Hysteria continued to exist as a psychiatric condition, but it eventually came to be known as hysterical neurosis; many of the other symptoms that used to be affiliated with hysteria are now affiliated with hormones, and they are described in terms such as "premenstrual tension" and "postpartum depression."

Picking up where previous historical analyses leave off, then, I explore in this book how even in the most recent scientific literature, experts are coming to terms with some of the fundamental changes in understanding that occurred as a result of the late nineteenth- and early twentieth-century discoveries that have received so much attention in the historical literature. Despite the fundamental changes in understanding that have resulted from the discovery of hormones, today's experts remain committed to a belief that the hormone-brain relationship in women's bodies is more difficult to control and understand than it is in men's bodies. These ideas are expressed in different ways; for example, they are expressed in neurologists' studies of hormonal influence on women's brain activity, in psychologists' studies of the behavior of female rodents at different phases of their reproductive cycles, and in psychiatrists' studies of mental health conditions in female patients.

Acknowledgments

This book was born in September 2013 after Lisa Meloncon had invited me, several months earlier, to give a keynote talk at the first University of Cincinnati's "Discourses of Health, Medicine, and Society" symposium. I had just finished a couple of major projects, and I needed a new one. Somehow, through the mysterious process that is rhetorical invention, *The Hormonal Woman* became that new project. Thanks to Lisa for giving me that exigency—without it, the book probably wouldn't exist. Thanks also to the many participants who listened to that talk and offered feedback that set the research and writing in motion.

I am also grateful to participants in the 2014 and 2015 Rocky Mountain Writers' retreats: Kaye Adkins, Lora Arduser, Dawn Armfield, Kristin Bivens, Tracy Bridgeford, Jennifer Brown, Jennifer Browne, Cassie Christopher, Kirsti Cole, Kenny Fountain, Robin Hart, Erin Justyna, Donna Kain, Amber Lancaster, Meghan McGuire, Kathy Northcut, and Candice Welhausen. The inspiration that came from hanging out and doing some serious writing with you all came at crucial times in the drafting process. Plus, it was a lot of fun.

My professional community at Texas Tech has provided a vibrant environment in which the material for this book could emerge and take shape. Colleagues and graduate students in Technical Communication and Rhetoric listened to me talk about this project at the May seminar in 2015 and 2016, and my graduate students in several iterations of the medical rhetoric seminar listened to me talk about bits and pieces of the research and offered insightful feedback. I especially want to thank Kelli Cargile Cook for intellectual inspiration, conversations about feminist matters, and serving as both a mentor and collaborator in the academic jungle. In 2015, Kelli also started a writers' group that brought together several of us and gave us space to discuss our research. Thanks also to Joyce Carter, Michael Faris, Becky Rickly, Abigail Selzer-King, Greg Wilson, and Rachel Wolford for participating in this group and taking their turns to host it. Chris Christofides at Texas Tech also deserves a huge amount of credit for stepping up at an extremely challenging moment in my administrative career that occurred right in the middle of drafting the manuscript. Chris settles for nothing less than perfection in everything she does. Plus, she makes me laugh on the days when that is most important.

Because of my January 2017 move from the College of Arts and Sciences to the College of Media and Communication, the circle of Texas Tech colleagues and friends I want to acknowledge has greatly expanded during the course of writing this book. I will be forever grateful to David D. Perlmutter, Dean of the College of Media and Communication, for making possible this new and completely unexpected professional opportunity. And every single day, I experience gratitude for all the faculty, staff, and students in the College of Media and Communication who have taken me in and made me feel welcome from the very first day.

Finally, this manuscript has benefitted greatly from presentation and networking opportunities provided by the Texas Tech University Faculty Research Club, the Texas Tech Humanities Center, and the Texas Tech Women Faculty Writing Program. The final round of revisions got a good jump-start at the Women Faculty Writing Program retreat in January 2017, sponsored by the President's Gender Equity Council and led by Elizabeth Sharp. And I was able to complete these revisions, even during an insanely busy time, because of this program's weekly three-hour writing sessions that were held throughout the course of spring semester 2017.

The project has benefitted from several forms of research funding at Texas Tech, including a Scholarship Catalyst Program grant that supported archival research in 2014–2015, a faculty development leave in 2015–2016, a Humanities Center Fellowship in 2016, and research funds from the Department of Communication Studies in 2017. Kristen Albert supported my research at the National Endocrine Society Sawin Library in May 2015, and Jack Eckert supported my research at Harvard University's Francis A. Countway Library of Medicine in July 2015. Florence Gillich and John Gallagher at Yale University's Harvey Cushing/John Hay Whitney Medical Library promptly and courteously helped me obtain reproduction-quality versions of the images of hysteria patients that appear in chapter 3. Thanks to the Cushing/Whitney Library for granting permission to reproduce these images. Thanks also to University of South Carolina Press for permission to include a copy of Lawrence Prelli's table in chapter 3.

I am grateful to everyone at Penn State University Press who has worked so hard to make this book possible. Kendra Boileau responded promptly and enthusiastically from the moment I contacted her to ask if the press might be interested. Series editors Leah M. Ceccarelli and Michael Bernard-Donals gave insightful and encouraging feedback from the beginning, and they always responded surprisingly quickly. The two anonymous reviewers also gave critical feedback that has greatly enhanced the quality of this manuscript, and the copy editing provided by E. A. Williams was superb.

Finally, of course, thanks to my family for putting up with my scholarly endeavors. It's just the nature of academic life that research and writing have to be fit in at the expense of time spent with family. I am so fortunate to have this job with a reasonably flexible schedule, but as my children have come to learn, a flexible schedule does not always mean tons of free time. I think we're doing OK, but I am aware of the sacrifice that everyone has to make for a family to thrive when both parents have demanding jobs. In marrying Brian Still, I know that I got the world's best spouse. And I know I have the world's three best children: Jack, Olivia, and Abe.

My extended family, although they are far away, also matter more than they realize. My mother, Gayle Backes, inspires me every day. I am proud of her and what she's accomplished. My father, Rodney Koerber, is probably responsible for my fascination with medical subject matter, even though I never wanted to deal with all the chemistry and math classes that would have been necessary to apply for medical school.

1

Hormones and Hysteria | A Rhetorical Topology

To capture the complexities of the long history through which hormones have come to replace hysteria, I invoke as a framework for this book's rhetorical history Michel Serres's theory of time as topological. In mathematics, topology is the "study of the properties that are preserved through deformations, twistings, and stretchings of objects."[1] The image that has come to be most closely affiliated with topology is the Möbius strip, as shown in figure 2. This geometric concept was named for the mathematician August Ferdinand Möbius, who was one of the two mathematicians working independently of each other who discovered this geometric concept in 1858. The distinguishing feature of the Möbius strip is that it possesses a continuous surface that remains smooth and intact no matter how the object is manipulated. Furthermore, the entire surface can be traveled perpetually, so that the whole surface space is covered without the traveler ever crossing a boundary or edge.

2 | Image of the Möbius strip.

Reflecting Serres's idea that the movement of time itself might be understood as topological rather than linear, this book's analysis unfolds from the assumption that underneath the smooth surface of a forward-moving, ever-progressing history of beliefs, there are undercurrents of backward movement and places where old ideas stand still. As another way to illustrate this understanding of time and historical progress, Serres draws an analogy to the River Seine, which seems on its surface to flow smoothly in one direction, even while beneath the surface water swirls about unpredictably and in multiple directions.[2]

In taking this approach, the book calls into question the assumption that our understanding of topics such as sex difference or "female problems" has advanced from an older myth-based model to a modern scientific model. Rather, throughout the long history that extends from ancient ideas about hysteria to modern ideas about hormones, a rhetorical-topological approach to this history allows us to see that every group of experts who has advocated new truths along the way has done so on the basis of a claim that they have access to more scientifically based, empirically grounded knowledge than did those who came before.

In fact, it is in Hippocrates's mechanistic explanation of hysteria that we see an early example of an expert who tried to refute the beliefs of those who came before him by offering a more scientific explanation of female problems. From today's perspective, of course, the Hippocratic treatises do not seem scientific. However, the rhetorical stance of these treatises was one in which the author was trying to assign natural, empirically observed causes of disease and to refute the religious explanations that preceded them. The prior religious explanations that Hippocrates was refuting assigned the cause of disease to supernatural forces such as demons. Thus, for example, the Hippocratic treatises include disparaging references to traditional healers who, instead of offering empirically based cures such as the use of odors, referred women to religious efforts such as dedications to the goddess Artemis. In the "empirical" understanding of women's health problems that Hippocrates offered in place of these older religious explanations, the wandering womb had an important role to play in almost all health conditions experienced by women.[3]

It is easy, from today's perspective, to dismiss these ancient views about the wandering womb as based purely in superstition. But even in these explanations that seem so far-fetched, this mythical or religious belief set cannot be seen as uniform, and from the perspective of the authors who wrote the texts, it was not based purely in superstition. Rather, the Hippocratic treatises claimed that their humoral theories of medicine were more empirically based than were earlier theories because they derived from symptoms that could be observed in the patient rather than from religious rituals intended to solicit help from the gods. And just as Hippocrates in the fifth century b.c.e. sought to dissociate his medical beliefs from the beliefs of those who came before him, the authors of each new generation's expert texts since then have sought to dissociate their beliefs from the beliefs of those who came before them. Since ancient times, I argue in this book, it has been a recurring pattern of argument for each generation's new set of experts to claim that their knowledge is more trustworthy, more scientific than that of the previous generation, and these experts have often used such claims to assert that their understanding of "female problems" can allow the women who suffer from these conditions to break free from the constraints imposed by the previous generation of experts.

Viewed through the lens of Serres's topology, the history of women's health that extends from ancient Greek times to the present is revealed as a history in which hormones have come to replace hysteria as both the motivator of women's behaviors and the root cause of many women's health problems. Along the way, there have been numerous reconceptualizations and transformations of the uterus, the brain, and the relationship between these two organs, and understanding these changes is integral to understanding this long process of historical change. Of course, we now have a greater number of scientific terms and verifiable facts that can be used to account for the phenomena we observe. But underlying these scientific terms and facts is a biological entity with a rhetorical function that has remained remarkably consistent through the centuries, whether it is labeled as a wandering womb, or as the out-of-control hormones that presumably govern women's health, behavior, and wellness at every stage of the life cycle in today's scientific and popular understandings.

In presenting and developing this argument, I echo and expand on some earlier feminist analyses of the history of hormones, which have often emphasized the revolutionary change effected by the discovery of hormones. For example, in the "hormonally constructed body" that emerged as a result of the early twentieth-century discovery of hormones, Nelly Oudshoorn argues, sex difference could no longer be traced to specific male and female organs, but rather was chemically infused throughout the bodies of men and women. Oudshoorn ultimately argues that "hormones have changed our world" by offering the promise that we can understand the chemical differences between men and women, and because these differences are chemical in nature, we can come to control them and thus improve on the "natural" body.[4] Emily Martin echoes Oudshoorn's observations, but Martin's focus is less on the implications for understanding sex difference and more on the role that hormones played in the rise of the machine metaphor of women's bodies that became prevalent in late nineteenth- and early twentieth-century medicine. As Martin says, in this mechanical understanding of the body, "the female brain-hormone-ovary system is usually described not as a feedback loop like a thermostat system, but as a hierarchy, in which the 'directions' or 'orders' of one element dominate."[5] Along similar lines, feminist scholar Celia Roberts connects the early twentieth-century establishment of gynecology to the discovery of hormones and, in particular, to the increasing availability of chemically synthesized hormones that brought physicians closer to their goal of greater control over women's bodies. Roberts also documents how the mid-twentieth-century discovery that men have some amount of estrogen in their blood and women have some amount of testosterone revolutionized expert understandings of sex difference by complicating earlier notions of hormones as "messengers of sex" that communicated a simple, straightforward, and deterministic message about the maleness or femaleness of a developing human body.[6]

More recent scholarship about the history of hormones continues to develop these insights, moving beyond sex difference to examine more closely how the rise of endocrinology has caused fundamental changes in our understandings of other aspects of human health. For instance, in her study of nineteenth-century discourses of infertility, Robin E. Jensen characterizes this era as a time during which human reproduction became increasingly medicalized. With the shift to an endocrine understanding of these processes, Jensen argues, these various aspects of reproduction came to be understood increasingly through a "chemical rhetoric of human reproduction."[7] This chemical rhetoric came to replace earlier understandings that were based in concepts such as "vital energy" and "moral physiology." As a result of this increasing faith in chemical explanations and solutions, "nature" came to be seen as "not something to be emulated but . . . remedied and perfected."[8] Along similar lines, Michael Pettit focuses on the new "glandular self" that emerged in the early twentieth century as a variety of commercial products started to be sold and advertised with the idea that stimulating the glands to produce hormones would have a rejuvenating effect that would combat the negative effects of aging and such. As public knowledge of the emerging field of endocrinology became more widely circulated in this period, Pettit contends, "widely publicized advances in endocrinology encouraged individuals to conceive of both their bodies and their fates as literally governed by a conglomeration of glands and their unseen chemical secretions." Pettit speaks of endocrinology as a "technology of the self" in that it allowed "personality and behavior" to be "understood as the *effect* of unseen chemicals."[9] He further describes this phenomenon as "the relocation of . . . social and economic struggles inside their own bodies, in the failure of their fluids to properly circulate."[10]

This book's analysis relies on many of these previous texts to develop and expand our understanding of the rhetorical, scientific, and historical implications of Dr. Ernest Henry Starling's coining of the term *hormone* in 1905. In fact, Serres's concept of topology is used to explore the history of hormones in chapter 1 of Roberts's *Messengers of Sex*,[11] and it is applied to the history of reproductive endocrinology in Jensen's work.[12] Whereas Roberts and Jensen each apply Serres's notion of topology to specific aspects of hormones' histories, in this book I apply topology more extensively to understand not only the rhetorical history of hormones, but also the long rhetorical history that preceded the emergence of the word *hormone* in 1905. In so doing, my analysis in this book aligns closely with that of Bernice Hausman, who argues that "in the [early twentieth-century] shift from ovaries to estrogen as *the* determinant of physiological and psychological femininity, we might say that the more things changed, the more they stayed the same."[13] Although Hausman does not deny that scientific progress has occurred in our understanding of sex difference, she suggests some important similarities between older and newer ways of thinking about scientific femininity. Thus, this book's stance is similar to Hausman's, and it also echoes and expands on some of Roberts's and Jensen's suggestions. However, none of these previous analyses has invoked Serres's topology to conduct a comprehensive analysis of the history of hormones, and none has closely examined the rhetorical history that connects hysteria to hormones.

Rhetorical Histories of Medicine

In rhetorical studies, our approach to historical topics in medicine has largely been shaped by Judy Segal's notion of kairology, which has proven to be a robust framework for illuminating connections among history, rhetoric, and culture. Kairology is based on the sophistic concept of kairos, defined by Segal as "the principle of contingency and fitness-to-situation."[14] Segal uses this concept to highlight the fleeting nature of truth in medicine and to overturn long-held assumptions that medicine is best understood as the technological application of ever-advancing scientific beliefs. Using examples that include hypochondria, migraines, and mental illness, Segal offers a useful framework for highlighting the ways in which society, culture, and science interact to produce a medical belief that is convincing at a distinct moment in time, and at a distinct geographic location, because it resonates with its intended audience at that place and time, for reasons that can include historical and social realities just as much as scientific ones.

As Segal says, "kairos is a way of making sense of both the medical past and the medical present."[15] She emphasizes how the "convergence" of themes or events at a given time and place leads us to practice and conceptualize medicine in a certain way. So, for instance, thinking in terms of kairology, in the twenty-first century, a rhetorical analyst might observe that we currently have an explosion of pharmaceutical products to treat mental-health conditions, and we also consequently have an explosion of new diagnoses that can be treated with such products. This growth in pharmaceuticals has also coincided with an increase in direct-to-consumer advertising and the easy availability of health information through the Internet. These events have occurred at the same time, in a mutually reinforcing manner, with a shift in the perception of the ideal physician-patient relationship. These various factors add up, from the perspective of kairology, to make it likely that the particular truths about mental health that circulate in today's world are perceived as accurate, universal, and timeless. This is Segal's rhetorical approach to the history of medicine, and subsequent scholars in medical rhetoric have employed versions of her kairology framework to explore the rhetorical histories of health-related phenomena such as breast cancer,[16] depression,[17] cancer treatment,[18] complementary-alternative medicine,[19] and pregnancy.[20]

Whereas Segal's kairology draws attention to the manner in which multiple factors in a rhetorical situation must converge in just the right way for a given medical belief or diagnosis to be accepted as valid at a certain time, Serres, by contrast, draws attention to the ways in which various elements that exist in science and the world at a given time and place do not add up, or do not make sense. Serres asserts that it is this dissonance—noise that interferes with clear communication—that leads ultimately to the production of new knowledge. As Steven D. Brown explains, one key focal point in Serres's work is the noise or interference that occurs when different disciplines attempt to communicate with each other: "It is in these spaces that confluences form as messages intersect with one another. The great inventive power of science is born in these confluences, where heterogeneous projects, social practices and ideas come together."[21] Through this emphasis on noise and conflict, Serres ultimately, according to Brown, offers us his distinctive approach to the history of science—an approach that eschews the notion that science advances through revolutionary breaks with past beliefs. Instead, Brown explains, Serres's philosophy suggests that science moves in many directions at once. He quotes the following passage from Serres's *A History of Scientific Thought: Elements of a History of Science*: "Far from tracing a linear development of continuous and cumulative knowledge or a sequence of sudden turning-points, discoveries, inventions and revolutions plunging a suddenly outmoded past instantly into oblivion, the history of science runs backwards and forwards over a complex network of paths which overlap and cross, forming nodes, peaks and crossroads, interchanges which bifurcate into two or several routes."[22] Through this emphasis on noise, confusion, and dissonance, Serres's ideas about scientific progress and change can help us broaden the approach of kairology that has been so influential in previous rhetorical histories of medicine. This broadening of our approach aligns with rhetorical theorists' recent efforts to expand our thinking about kairos to include more than just proper timing or decorum. Thomas Rickert, among others, has called for this expansion, urging us to account for the embodied, spatial, and material aspects of kairos that are obscured if we consider only proper timing and keen rhetorical judgment.[23] Along similar lines, Michelle Ballif challenges rhetorical scholars to revise our approaches to historiography to account for the extent to which "'history' *is* always already 'out of joint,' out of time, untimely, anachronistic"[24]—another way of asking us to think beyond the emphasis on kairos or timeliness that has grounded so much of our historical scholarship in rhetorical studies.

Looking more broadly at recent scholarship in the rhetoric of science, the topological approach to rhetorical history that I employ contributes also to conversations about incommensurability, rhetorical invention, and scientific progress. Scholars such as Randy Allen Harris,[25] John Angus Campbell,[26] and Celeste Condit[27] have made clear that the rhetorical tradition gives us a rich set of tools for understanding how the new emerges from the old and, thus, for reminding us that novel ideas, even in science, are usually not quite as new as they are made to seem. This book's topological approach advances these conversations by expanding our vocabulary for describing the precise ways in which the new emerges from old, and the ways in which new ideas preserve elements from the old.

In addition to advancing these theoretical and methodological conversations in rhetorical theory and the history of science and medicine, another important dimension of the book's analysis is the subject matter itself. Examining the history of hysteria and hormones forces rhetorical scholars to confront the realities of the sexed body to an extent that simply does not happen when our subject matter remains limited to Charles Darwin's texts, the discovery of DNA, and the like. Although the attention that rhetorical theorists have paid to subjects like these has led to important insights about the fundamentally rhetorical nature of scientific thought, it has left intact some central assumptions about what counts as knowledge that is worth investigating. As this book's analysis will make clear, when we take hysteria as a starting place, many of these assumptions are, from the beginning, undermined. Rather than looking only for the great moments of discovery, the times when individual great male thinkers suddenly arrived at brilliant insights or new truths, hysteria forces us to acknowledge that darkness, complexity, and puzzlement are just as important as clarity, light, and insight in the overall knowledge production endeavor. This point is emphasized in Serres's efforts toward a fundamental rethinking of our Western approach to the knowledge endeavor. Rather than emphasizing enlightenment or illumination, he says, we might also consider darkness, or bringing together disparate strands. This technique would, of course, go against the grain of the perpetual desire in Western intellectual tradition to clarify and separate our objects of analysis from each other: "We should invent a theory of obscure, confused, dark, nonevident knowledge—a theory of 'adelo-knowledge.' This lovely adjective, with feminine resonances, means something that is hidden and does not reveal itself. The Greek island of Delos was once called Adelos, the hidden one. If you have tried to approach it, you surely know that it is usually hidden in clouds and fog. Shadow accompanies light just as antimatter accompanies matter."[28] Much of the scientific work that has been done on sex difference throughout the centuries is work that is devoted to just this kind of classifying, separating, conceptualizing, clarifying, illuminating, and measuring that Serres asks us to reject. For the most part, rhetorical theorists who have taken scientific argument as their subject matter have upheld this tendency to idealize the big moments or breakthroughs, even as they have illuminated the previously unseen rhetorical dimensions of scientific change. By contrast, rhetorical-topological analysis reveals that even as scientists in multiple disciplines tried to make sense of sex difference through intense efforts to develop sex- and race-based classifications of brains, behaviors, and bodies, these classifications and hierarchies repeatedly collapsed on themselves as researchers failed to deliver the empirical evidence that was increasingly demanded to support their claims in a climate in which medicine as a disciplinary establishment was trying to become increasingly affiliated with modern science rather than the ancient beliefs that had been passed down for generations. It is this failure to deliver acceptable evidence that created the exigency against which, eventually, hormones were able to emerge as the acceptable scientific explanation to replace old beliefs such as the wandering womb that had persisted, in various manifestations, since ancient times. Hormones seemed to cast clarity and light on a domain that for centuries had been shrouded in darkness, myth, and obscurity. In this book, I use rhetorical-topological analysis to explore the light that was shed by hormones, and to acknowledge the new forms of darkness, myth, and obscurity that we often ignore because of our post-Enlightenment desire to illuminate, clarify, categorize, and separate.

In short, I argue, previous approaches to rhetorical history in science and medicine can be usefully enhanced with Serres's concept of time as topological—which, as explained by Serres, entails a conception of history as folding in on itself in a manner that resembles the kneading of dough. As I demonstrate in subsequent chapters, Serres's concept of time both complements and problematizes previous understandings of rhetorical history and change in the context of health and medicine by exposing the precise manner in which old expert-based or religious beliefs can resurface, even after they seem to have long ago been disproven or forgotten. Taken together, these theoretical concepts point the way toward a distinctly rhetorical approach to the history of science and medicine— an approach that is especially well suited to the realities of the sexed body that have been largely ignored in previous rhetorical scholarship. At the same time, through this analysis, I demonstrate how early endocrinologists preserved key elements of hysteria. They did so indirectly, through a complex web of "backward," "sideways," and "forward" rhetorical moves that occurred surreptitiously, even while they made a conscious effort to use new vocabularies and epistemological frameworks.

To take this topological approach, of course, does not mean eschewing altogether the value of science, nor does it imply that we should go back to old ideas like that of witchcraft. As Condit has recently reminded us, even as we critique science, rhetorical scholars must avoid unintentionally aligning ourselves with those contemporary "political regressives" or the "warlords" who would gladly dismiss most scientific endeavors, except for those that support their efforts "to eviscerate what we have come to know as the large-scale state in order to expand their personal power." In Condit's words, we must not "portray science as an evil enemy," and we must not "refuse to be held accountable for the impressive and important bodies of evidence that natural and social scientists have amassed." We can, however, insist that "scientists cannot achieve what they claim to want to achieve—maximally inter-subjective, trans-situational descriptions—unless they are also self-reflective about the role of language in their endeavors."[29] And this means continually asking what else, in addition to scientific progress, has been involved in this dramatic shift in the framework for understanding women's mental and physical health that is this book's focus. What are the rhetorical moves, events, and phenomena that have been involved in the dramatic shift in thinking about women's health that has obviously occurred between ancient Greece and the present? What are the forces, both cultural and scientific, that have led to this apparent change in beliefs? And—specifically of interest for this book—what role has been played by the rhetoric and science of hormones in this long history?

Historical Scope and Research Methods

The emphasis on kairos has infused much recent scholarship in medical rhetoric, and the nineteenth century is often singled out by scholars as the era in which medicine came to be increasingly tied to science rather than to mythical or religious beliefs. The nineteenth century is the era of Foucault's "clinical gaze" and the "birth of the clinic,"[30] and in Segal's kairology of biomedicine, the new body that emerged in this era was "hardbound, anatomized, and deferent to the institution."[31] This newly conceived body, subject to the clinical gaze more than ever before, replaced the "humoral" body of previous eras. Reflecting this view of history, medical rhetoric scholars have often focused on the nineteenth century as a major turning point at which a wide array of health conditions—including everything from migraines, to childbirth, to sexuality—became medicalized.

Because of everything that was occurring during the nineteenth century, with the increasingly intense effort to understand male-female difference in scientific terms, it may be tempting to begin this book's historical analysis with the nineteenth century. Of course, this era is widely known as the time when scientific explanations started to replace religion in virtually all aspects of life. Starting with the nineteenth century also seems logical because it is the period that immediately precedes Starling's 1905 lecture in which he coined the term *hormone*. However, an important contention of this book is that focusing too much on the nineteenth century as a distinct point in time makes it easy to lose sight of trends that actually began much earlier. In fact, echoing Serres's observations about scientific progress and time as topological, a closer look at everything that was occurring in nineteenth-century science and in the broader public sphere makes clear that experts' claims to understand women's health problems more scientifically than their predecessors did actually extends to ancient texts. Thus, the scope of this book is quite large, extending from ancient texts to the present. By traditional standards of historical analysis, this scope might seem too large. But as I hope to make clear, taking a broad scope is beneficial, even necessary, because of the extent to which ancient ideas about women persist in today's science and popular culture.

The research for this book has involved extensive rhetorical analysis of texts, and it reflects Kelly Happe's observation that "a rhetorical analysis attends not only to shared beliefs across multiple discourses, but also to the inner workings of the texts that form them."[32] The texts that I have consulted include secondary historical analyses of the older texts as well as primary texts for coverage of the more recent centuries. On the subject of hysteria, historians have documented the mythical and scientific beliefs about this disease that have persisted between ancient times and modern, and they have noted the great confusion that has surrounded this disorder, even up through the late nineteenth and early twentieth centuries when hysteria slowly started to slide out of fashion as a medical diagnosis.[33] Feminist analyses have demonstrated how cultural beliefs, such as widespread mistrust of women's bodies, have made it virtually impossible to conduct objective scientific research about hysteria, and have highlighted the ways in which hysteria diagnoses and treatments have been used to regulate women's behaviors and to exhort individual women to adhere to social ideals in hopes of restoring the social order at times when that was threatened.[34]

Many of the primary texts that document the history of hormones have been easy to access online. For instance, Starling's 1905 lectures are available in a digital version that has been reproduced from the Bodleian Libraries, and this same online resource includes several additional texts that help to situate my claims about the lectures' rhetorical history. The National Endocrine Society's organizational website also offers access to a series of oral history videos and an online archive that includes 8,100 texts from the Sawin Library's collection. These online resources have been useful in illuminating the history of hormones, and of endocrinology as a field, but I also traveled to Washington, D.C. in May 2015 and spent three days researching onsite in the Sawin Library. In July 2015, I spent two days researching hysteria at Harvard University's Countway Library of Medicine. The resource that I consulted was a textual archive that included patient charts, handwritten notes, and photographs from the Salpêtrière Hospital, where Dr. Jean-Martin Charcot treated hundreds of hysteria patients in the late nineteenth century. I have followed up on this archival research with access to the digital collection of photographs from the Salpêtrière Hospital, which is available through Yale University's Harvey Cushing / John Hay Whitney Library.

Structure of the Book

The book is divided into sections that reflect the major themes that extend across the centuries as experts in various generations have attempted to define what makes women who they are and to solve the female problems that they have experienced since the beginning of time. Similar to the way in which, according to Serres, the Seine River has a smooth surface that seems to flow in one direction, the book's overarching structure is based on three chronological periods: the period that spans several centuries prior to the discovery of hormones, the period immediately preceding and following the discovery of hormones, and the period after the discovery of hormones. This overarching structure is akin to the smooth surface of the Seine, which appears to move only in one direction. Within the three broadly defined sections, though, the chapters are structured in ways that are akin to the undercurrents and confusion that lie underneath the river's smooth surface. Thus, within the chapters of each section, the book's analyses are organized around major themes that extend through the centuries. Sometimes these themes disappear for a while and then resurface, so the analyses within these chapters are presented in a way that intentionally captures the complexity and nonlinear form of emergence and evolution of scientific ideas over the centuries, and the multidirectional transfer of knowledge from expert to public spheres and back again. The scientific change that has occurred while calendar time has progressed can be understood, thus, as a process of rhetorical change, in which new scientific beliefs continually emerge as the language of each era has to be morphed, twisted, and reshaped to fit the demands of each new moment.

In addition to these efforts to complicate our understandings of the progression of time in scientific thought, each chapter also complicates specific rhetorical concepts that have been used and adapted through the centuries of rhetorical history. These concepts include topos (chapter 2), stasis (chapters 3 and 4), memory (chapter 5), metaphor (chapter 6), and enthymeme (chapter 7). Although each chapter invokes a specific rhetorical concept, this structure is not meant to suggest that only one rhetorical concept was operative at any given time or in any set of texts being analyzed. The chapter structure does not imply, for instance, that topos was no longer relevant after the ancient texts that are the main focus of chapter 2. Rather, just as the linear progression of history that might seem to be suggested by the book's overall chronological organization overlays a set of themes that continue to swirl around unpredictably underneath the surface structure of the book, these rhetorical concepts overlay an unwieldy set of rhetorical patterns that move, almost of their own accord, beneath the surface of a chapter structure that outwardly displays a primary focus on only one rhetorical concept at a time in order to provide readers with fleeting moments of stillness amid a book that is fundamentally about movement. In fact, an important additional purpose of the book itself is to understand rhetoric as movement. Whether we are talking about advertising language, scientific discourse, instructional language, or some other form of rhetoric, we can define rhetoric as the use of language or symbols in the hopes of effecting some kind of movement. As I argue throughout the book, there are multiple ways to see topological dynamics in science as moving neither backward nor forward, but simply moving—that is, as changing, sometimes significantly, while also staying the same. Using a specific rhetorical concept for each chapter's focus enhances our understanding of this phenomenon by helping us visualize the precise form of movement that is most salient to the kind of rhetorical activity that occurs in the particular set of texts that each chapter analyzes. In short, each of the rhetorical concepts (topos, stasis, memory, metaphor, and enthymeme) that I employ in the analysis chapters is a concept that rhetorical theorists have used—extending from classical to contemporary times—to illuminate some aspect of the many ways in which ideas move. This movement of ideas can entail movement across space or time, movement in speakers' and audiences' beliefs or mindsets, or even a temporary lack of movement (as in the case of stasis). Thus, I foreground different analytic concepts in each chapter because the series of events that are highlighted in each chapter demand different ways of understanding rhetoric as movement. In terms of research or analytic method, these rhetorical concepts were not imposed on the texts in a top-down manner; rather, they emerged from the texts through my own efforts as a rhetorical analyst who chose to listen critically as I analyzed the textual artifacts that constitute each chapter's focus and the conversations occurring in the artifacts to discern which rhetorical concept would be most appropriate to help us understand the specific form of movement that was unfolding at that historical moment in that particular set of texts. In each chapter, then, I employ a specific rhetorical concept to guide the analysis, but in so doing, I also consider how the concept is reconfigured or called into question by the texts that are analyzed in the chapter and by a specific element or concept from Serres's work on topology.

In chapter 2, I explore the early history of hysteria, beginning with ancient medical texts from Egypt, China, and Greece and tracing how some of these ancient ideas persisted, quite literally, in texts that were best-sellers among nineteenth-century readers in the United States and in Europe. The chapter's analysis uses this expansive coverage of hysteria's early trajectory to suggest the major themes and patterns that are revealed by a rhetorical-topological approach to the history of medicine. In particular, the chapter pairs Serres's ideas about novelty in science with rhetorical scholars' recent efforts to rethink classical theories of invention. Taken together, these theoretical concepts illuminate the womb as a topological space that has been consistently present—yet has undergone numerous changes—throughout the centuries-long history of hysteria.

Chapter 3 continues exploring hysteria, taking a closer look at the fragmentation that started to occur in the nineteenth century as the discipline of medicine itself became increasingly specialized into the numerous subdisciplines that were emerging at this time. Relevant subdisciplines that are explored in this chapter include neurology, psychology, neuropsychiatry, physiology, and endocrinology. As a result of this fragmentation of medicine, and of the human body more generally, hysteria itself was divided into numerous conditions, each understood in a different way by the various disciplines. The chapter invokes the rhetorical concept of stasis to characterize scientific arguments about hysteria in the late nineteenth century, using Serres's observations about noise, conflict, and silence to add to rhetorical scholars' understanding of the many contradictions that can exist within moments of stasis.

Chapter 4 explores how the increasing fragmentation of nineteenth-century science, along with the intensification of other divides such as that between science and religion, provided an ideal context for a concept such as hormones to emerge in the early twentieth century. The chapter closely examines the 1905 lecture in which Starling coined the term *hormone* and exposes key features of the rhetorical situation that immediately preceded and followed the coining of this term. Stasis is again the key rhetorical concept in this chapter's analysis. Whereas the previous chapter emphasizes the stillness or lack of movement that exists in a moment of stasis, this chapter highlights the momentum that exists in stasis and explores what happens when that momentum is released.

Chapter 5 examines how the general concept of hormones came to be shaped more specifically into types such as sex hormones and, even more specifically, female and male hormones. Through close analysis of scientific texts published in the first half of the twentieth century, this chapter exposes how "female sex hormones" came to be understood in increasingly precise terms that allowed scientists to move away from vague notions about the uterus, the female brain, or maternal instinct. Through this analysis, the scientific discourse of sex difference can be understood as a discourse of remembering and forgetting. In this discourse, hormones became a successful scientific concept because they enabled old understandings of femininity to be communicated in ways that would resonate with audiences who had increasing expectations that scientific evidence would support the beliefs that they would accept.

Chapter 6 explores how the hormonal understanding of women's bodies that was advancing in the early twentieth century enabled experts to replace the vague and confusing notion of hysteria with a hormonal explanation for many of the symptoms and behaviors that had, for centuries, been affiliated with hysteria. The chapter's focus is the emergence of diagnoses such as premenstrual syndrome that have gradually replaced hysteria over the course of the twentieth century and into the early twenty-first century. With the transition from the hysterical-woman metaphor to the hormonal-woman metaphor, chapter 6 reveals, physicians gained a more robust vocabulary to describe the multiple ways in which women's bodies can be pathologized. In rhetorical terms, I argue in chapter 6, hormones served as the metaphorical concept that allowed ancient ideas about women's bodies to be transported into the twentieth century.

Chapter 7 continues to explore the most recent scientific and popular discourses but with a focus on women's mental capacity rather than on mental illness diagnoses. The chapter explores how conditions such as "pregnancy brain" are studied in the scientific literature and how they are formulated in popular discourse. As I argue in this chapter, the very fact that scientific experts continue to investigate a subject such as the effect of pregnancy hormones on women's brains perpetuates a long-standing assumption that women need to concern themselves with their body-brain relationship, whereas for men this relationship is assumed to be seamless and unproblematic. For each body of research that the chapter examines, I expose what has been omitted and how this omission serves as an enthymeme that causes the argument to resonate with specific audiences.

Chapter 8 concludes the book and suggests directions for further research. As new technologies enhance our ability to visualize brain activity, I argue, the scientific fascination with the brain-hormone relationship will continue to demand critical attention from scholars of feminist rhetoric. The topic also has broader public significance because, in popular texts, these rhetorical configurations of brains and hormones are often communicated in ways that make them seem progressive or empowering to women. In addition to summarizing the book's contributions to scholarly conversations, chapter 8 draws connections to current public debates about topics such as reproductive justice, environmental issues, workplace diversity, and the politics of academic knowledge production.

As these chapters demonstrate, we have obviously seen many changes in our understanding of women's biology, physiology, psychology, and general health during the centuries that have unfolded between the time of Hippocrates's medical treatises and the present. On the surface, these changes appear to have accomplished a complete dismantling of the mythical and religious beliefs that provided the framework for our understanding of women's health throughout much of recorded history. That is, most of us would like to believe that today's beliefs about women's physical and mental health—beliefs that are far more likely to attribute health problems to hormonal problems or chemical imbalances than to a wandering womb—are based in an entirely different framework of beliefs than were their ancient predecessors, and that this framework is far superior to that which informed the ancient ideas. Looking underneath this smooth surface, however, reveals a far more interesting story in which scientific "progress" proves to be anything but straightforward.

2

Hysteria from Ancient Texts Until the Nineteenth Century | The Womb as Topological Space

When scholars study the history of specific health conditions, they can usually identify a specific moment when uncertainty and guesswork gave way to scientific explanations and treatments that are proven to have some degree of efficacy. So, for instance, when it became possible to produce and market insulin as a drug, diabetes was transformed from a certain death sentence to a chronic condition that could be managed and lived with for many years.[1] When pharmaceutical products became available to treat migraines effectively, migraines came to be understood as a specific type of headache rather than a personality defect or a consequence of unhealthy behavior.[2] When medications proved effective at lowering blood pressure, hypertension came to be understood as a long-term or chronic condition that is relatively benign as long as the patient complies with the physician's recommendations.[3]

This phenomenon of new discoveries or breakthroughs in science and medicine is one of the most central forms of movement that rhetorical scholars have studied and, of course, such scholarship has produced significant evidence to complicate the notion that significant changes like these occur at a single moment, or as the response to a single discovery of a lone scientist working in the laboratory. By studying science as argument—and by applying rhetorical concepts such as topos and kairos—we know now that discovery is a much more complex form of movement than might have been previously believed, that no new idea is ever entirely new, and that the impression of scientific advancement as the product of new facts that erupt fully formed at a single moment is usually itself a product of carefully executed rhetorical activity on the part of individual scientists or research teams. Nonetheless, for most of the objects of study that have received significant attention from scholars of the history of medicine—including diseases such as diabetes, migraines, and hypertension—it is possible to identify a turning point—a place and time—when it at least appears to be the case that the disease being studied that was previously seen as incurable came to be understood as a condition that might be chronic but could be managed through the proper application of techniques or substances developed by bio-medical scientists.

In this regard, hysteria is a health condition that stands in sharp contrast to most others that have been extensively studied by historians of medicine. Never has a health condition caused so much confusion for medical experts yet, at the same time, been subject to so many different interventions and treatment recommendations. As stated in the previous chapter, medical use of the term *hysteria* can be traced from the texts of Ancient Greek physicians up until 1994, when the term *hysterical neurosis* was finally removed from the fourth edition of the *Diagnostic and Statistical Manual.*[4] However, unlike the many other health conditions that have been studied by historical researchers, we cannot identify a moment when it appears from the outside that hysteria came to be understood more clearly or to be seen as a treatable condition. Rather, it seems that the harder that experts worked to find answers, the more this disease eluded them. Thus, in the case of hysteria, confusion and conflicting recommendations intensified over the centuries until finally, sometime during the twentieth century, it became fashionable to move on to other ways of explaining the multiple symptoms that had come to be affiliated with this primarily female condition.[5] As stated in a 1965 article in the *British Journal of Medicine,* "it is generally agreed that no one has yet framed a satisfactory definition of 'hysteria'; but it is usually claimed that it can be recognized when met with."[6]

Through close analysis of medical texts published between ancient times and the nineteenth century, this chapter exposes the many ways in which hysteria eluded even the best and brightest scientific and medical experts. This chapter's close analysis reveals how the meaning of hysteria has changed throughout the centuries as each new generation of experts has attempted, yet failed, to craft explanations more scientific than those of their predecessors. Drawing from Serres's notion of topology, I argue in this chapter that the womb has persisted as a key construct, and throughout the centuries this female organ has been continually reshaped to meet the needs of each new generation's experts. As these different shapes of the womb have evolved and emerged, they have preserved continuity with past shapes, stretching and twisting without bending or breaking, and thus resembling the topological space of the Möbius strip.

In addition to enhancing our historical understanding of this disease, applying the notion of topological spaces to the history of this mysterious female condition also advances rhetorical scholars' understanding of scientific invention and discovery as forms of movement that are characterized by a pattern that is less predictable than we might see in rhetorical histories of other diseases and health conditions. This unpredictability is captured in some of Serres's insights into the backward and forward motions that, he observes, lie underneath the surface of scientific progress. As explained by Bruno Latour, most French philosophers have a proclivity for revolutionary ideas of the sort that reject everything prior and pronounce broad-sweeping intellectual change. But Serres's philosophical project works against this trend and emphasizes instead that revolution, or significant changes in belief systems, are never quite as broad-sweeping as they seem to be on the surface: "Every time a revolution . . . has definitely reversed the order of things, [Serres] still believes in what has been reversed; worse, he does not know how to choose between the past and the present, the losers and the winners."[7] Furthermore, says Latour, "the past of science, for Serres, is still active. No revolution in physics has covered up the Epicurian approach of fluctuations, no more than the invention of the genre of scientific writing has disactivated mythology, cosmogony, foundation stories or fables."[8] Latour's interpretation of Serres's philosophy offers a useful starting place, but hearing Serres's response to Latour's interpretation is also important. In an interview with Latour, Serres makes clear that despite his suspicions of the most revolutionary knowledge claims, he does not completely deny the possibility of new knowledge production in science. Rather, he emphasizes a paradox inherent in scientific discovery—namely, that one cannot articulate new ideas without depending, in some way, on the old ideas that have come to be accepted as truth. Serres explains this paradox in his interview with Latour: "It's much harder than we think to guard against accepted ideas, because often the ideas that seem the most modern, that suddenly mobilize a whole community—its media and its conversations—are agreed-upon ideas. In order for an idea to circulate it needs to be polished; it always takes years for it to acquire that smooth surface that enables it to circulate. This is why the ideas that circulate are usually astonishingly old. Thus, he who seeks newness remains alone."[9] In this emphasis on the interconnectedness of new and old ideas, we can see how Serres's efforts to rethink traditional assumptions about scientific knowledge production offer a useful scaffolding for rhetorical interpretation of a subject such as hysteria. Rather than understanding scientific advancement as a largely forward form of movement, Serres prepares the way toward a view of invention and discovery as both moving in fits and starts and following patterns that can be quite unpredictable.

As Carolyn Miller observes, in Greek and Latin the words for the verb *invent* "ambiguously include what are now two senses: that of coming upon what already exists (discovery) and that of contriving something that never existed before (invention)." In the way that we typically use the word *invention* now in English, Miller contends, we have lost that dual meaning, so "invention, especially outside the domain of rhetoric, has come to concern novelty." This is an interesting contrast to the rhetorical tradition, however, because "in rhetoric . . . the former sense has traditionally been assumed: the rhetor examines a preexisting inventory of 'stock arguments' and 'commonplaces' to select those that are most appropriate to the situation at hand."[10] Miller says that this divide between the two meanings of invention—with the emphasis placed on novelty in the latter meaning—originated in the sixteenth century and then became further entrenched with Francis Bacon, who distinguished between two kinds of invention: the kind that occurs in the arts and sciences, and the kind that occurs in speech and argument. In contrast to this long-standing perceived divide between the different forms of invention, Miller urges rhetorical scholars to return to older meanings that emphasize both dimensions of *topoi*. In so doing, she offers an understanding that resonates in some important ways with Serres's reflections about the relationship between old and new in scientific knowledge production.

Serres's notion of topology emphasizes smooth, continuous surfaces and continuity of ideas, but rhetorical concepts of invention are a useful counterpart because they capture precisely how change—including movement that is sometimes in a forward direction—can occur even on this smooth, continuous space. If the surfaces were entirely smooth and continuous, then scientific change would have to be perceived as occurring forever in a closed loop. In terms of Serres's image of the Möbius strip, for instance, it is hard to explain change or acknowledge any degree of scientific progress, if movement can only occur along that closed loop. But when we acknowledge the multiple dimensions of rhetorical invention, we gain a clearer picture of how new ideas can emerge from the material of old ideas. No idea is ever entirely new, nor is it entirely old. Although classical theories of invention usefully complement Serres's ideas, though, I argue in this chapter that Serres's ideas about knowledge production—particularly his notion of "adelo-knowledge"[11] and related comments about knowledge and femininity—offer an essential complement to classical rhetoric when it comes to understanding the history of a disease like hysteria, which has been so persistently implicated in scientific and medical concepts of the female-sexed body.

Employing these ideas as a theoretical framework, this chapter's specific focus is a collection of texts that circulated widely among nineteenth-century audiences both in England and in the United States. The first is *Aristotle's Masterpiece*, which was one of the most popular sex manuals available in the mid-nineteenth century and a text that had been an important source of popular information for several centuries prior. The first English edition of *Aristotle's Masterpiece* was published in 1684,[12] and Latin versions have been traced to the early medieval period.[13] In the eighteenth century, this text apparently became the "best-selling guide to pregnancy and childbirth"[14] and was published in a greater number of editions than any other similar book available at the time. The popularity of this sex manual continued into at least the early nineteenth century, and it was still being sold in the United States during the early twentieth century in a version that had remained mostly unchanged since the eighteenth century.[15] Although the text mimics many of the stylistic features of Aristotle's other texts, Aristotle was clearly not the actual author. In fact, the 1846 edition included contemporary examples and references to Biblical figures and Christianity. The 1846 edition is subtitled *The Midwife's Guide*, and it includes much information and advice specific to midwifery as practiced in this era. Aristotle's text offered comprehensive coverage of everything related to male and female reproductive organs, as well as the processes of conception, gestation, and delivery. This text was continually reprinted and circulated, and it offers extensive evidence to illustrate the extent to which ancient beliefs about conditions such as the wandering womb still formed an important part of the popular imagination about women's health in the mid-nineteenth century.

The second object of analysis is a collection of texts by Dr. Fredrick Hollick. Hollick was a prolific author who published several books about popular health in the nineteenth century, but two books in particular provide the focus for this chapter's analysis. The first book, published in 1847, is titled *The Diseases of Woman: Their Causes and Cure Familiarly Explained.*[16] The second was published in 1902 and is a collection of lectures that Hollick had delivered during several decades prior. This nearly 1,000 page book is titled *The Origin of Life and Process of Reproduction in Plants and Animals.*[17] In this book, Hollick explains everything from the origins of life to plant reproduction and recommended behaviors for men and women in marriage. With chapter titles such as "Influence of the Brain over Generative Powers (Chapter XLL)" and "Displacement or Wrong Position of the Female Organs (Chapter LIII)," this manuscript provides excellent insights into popular and scientific understandings of the brain-uterus relationship just before the term *hormone* was coined in 1905 and a couple of decades before estrogen and progesterone were identified as the hormones that cause menstruation. The content of the two books overlaps somewhat, but as the title suggests, the first focuses more directly on women's health, whereas the second offers more comprehensive coverage of the health problems that both men and women could experience.

Coinciding and Conflicting Explanations of Hysteria

Inspired by Serres's idea of adelo-knowledge, this analysis of past beliefs about hysteria takes a different approach than might be expected. Rather than attempting to demystify or illuminate hysteria, this analysis foregrounds the confusion and conflict that permeated experts' understandings from ancient times through the nineteenth century and traces how these confusing and conflicting ideas persisted in the nineteenth-century texts that constitute this chapter's focus. Although each new explanation of hysteria that is introduced in the analysis below can be traced to a distinct moment when it seemed to appear, the overall development of ideas reflects Serres's understanding of time as topological in that newer ideas do not necessarily replace the older ideas in a progressive fashion. Thus, the analysis in this section is organized around three different explanations of the womb's role in hysteria—explanations that cause this female organ to be continually reshaped, morphed, and twisted to meet the demands of each new group of experts in a manner that leads the womb itself to possess topological characteristics. Although these different shapes of the womb directly contradict each other to some extent, they can be traced from the most ancient texts to nineteenth-century texts, in which we see them coexisting and reflecting the great confusion and tension among modern science and older ways of knowing that characterized intellectual life in the nineteenth century. First among these three explanations is the physical explanation of the womb as a cause of hysteria. The second is the humoral explanation. And the third is the moral explanation.

Physical Explanation of the Womb

The physical explanation of hysteria rests on the idea that some kind of irregular movement of the uterus caused this organ to be displaced and thus to interfere with other functions in women's bodies. This is one of the oldest medical explanations for hysteria, with roots that can be traced to ancient Egyptian texts, ancient Chinese texts, and ancient Greek texts.[18] The earliest origins of the wandering-womb concept have been traced to textual fragments from Egyptian papyri that date as far back as 2000 B.C.E. The idea that the womb caused health problems by moving around in women's bodies is a prominent and recurring theme in these papyri. In some places in these texts, the womb's wandering is said to have occurred when the womb suffered from starvation. In other words, it is suggested that having fed the "hungry" womb with a pregnancy would have resolved the distressing symptoms that caused it to move. These ancient texts also offer a variety of treatments other than pregnancy, most of which depended on using substances with pleasant or unpleasant odors—placed either at women's mouths or vaginal openings—to attract or repel the womb so that it would return to its natural location.[19]

Historian Beng-Young Ng also identifies in ancient Chinese medical texts evidence of conditions that are similar to hysteria. These conditions are first documented in a text believed to have been written around 200 C.E. by Zhang Zhong-jing, who is sometimes referred to as "the Hippocrates of China."[20] The conditions resembling hysteria have names such as "hurrying pig illness" and "sickness of the hasty organ" in these ancient Chinese texts. Such names obviously imply that something in the body is moving rapidly and causing disruption, and many of the symptoms that this author mentions are symptoms that would now be affiliated with anxiety. Reflecting the epistemological framework of Chinese medicine, the conditions and symptoms are described in terms of warmth and coldness in different parts of the body. According to Ng, it is documented in one of these ancient texts that "if a menstruating woman caught the 'cold injury,' the uterus became invaded by 'heat,' and she would be clear in mind during the day but would become talkative at night as if she had met a ghost. This was due to the heat entering the blood chambers. It was believed that the eventual discharge of heat through the menstrual flow augured a good prognosis."[21] Ng describes a condition called *zang-zao* that is described in a different ancient Chinese text; the woman who would have suffered from this condition would have been "easily saddened and crie[d] often" as well as "emotionally unstable as if influenced by an evil spirit, yawn[ing] and stretch[ing] her limbs frequently."[22] He notes that Chinese medicine is organ-based—each of the internal organs is understood as housing a particular kind of energy or emotion—and he provides another quotation from the same text to demonstrate that the hysteria-like condition known as *zang-zao* originated in the uterus: "Suddenly, the patient faints and loses consciousness, or becomes worrisome and sad, and bursts into temper without reason; this is due to disease of the reproductive organ, and not really related to devils and spirits."[23] He documents yet another condition, *ben-zhu qi*—which translates to the "hurrying pig illness"—that he says "occurs like an attack, beginning with the sensation of air arising from the lower abdomen to the chest, then rushing up through the larynx."[24]

In the Western tradition, Plato and Hippocrates are cited as the earliest sources of the idea that the womb could physically move around in the body. Plato's (428–347 B.C.E.) concept of the womb was animalistic; he refers to the womb as the "animal within" women that was "desirous of procreating children," and he suggests that when this desire was unfulfilled, the womb began to wander uncontrollably throughout the body, causing various health problems.[25] In contrast to Plato's animalistic concept of the womb, Hippocrates (460–370 B.C.E.) postulated a more "mechanistic" notion of the womb by suggesting that when the womb lacked the "moisture and fullness" obtained through pregnancy or intercourse, it "was attracted by sources of moisture," and this desire for moisture was what led the womb to engage in its wanderings. Regardless of these differences between Plato's and Hippocrates's concepts of the womb, though, both authors echo the recommendation from ancient Egyptian texts that the condition of the wandering womb could be resolved through intercourse and, hopefully, a resulting pregnancy.

It is not entirely clear, based on historical evidence, whether Plato's and Hippocrates's concepts of the womb served as competing explanations for hysteria, or whether they might be better understood as complementary understandings that originate in two different disciplinary perspectives. Although some historians of medicine suggest that the notion of the womb as wild animal was once accepted as literal truth, recent historians have suggested that Plato's use of this language was never intended literally, as an opposing school of thought to that being offered by physicians such as Hippocrates.[26] Rather, these historians suggest, the differences between Hippocrates's and Plato's beliefs are essentially disciplinary differences—that is, whereas Hippocrates's mechanical explanation was included in a medical treatise, Plato's animalistic explanation was meant to be poetic or descriptive, and according to these historians, it can be best understood as a literary derivation from Hippocrates's ostensibly scientific ideas. Specifically, Plato's description "embodies some ideas possibly based on Hippocratic discussions about the suffocation of the womb but greatly embellishes the original medical texts."[27] Nonetheless, subsequent authors continued to engage with Plato's ideas, even if they did not agree that the womb was literally an animal. For instance, Soranus of Ephesus, who practiced medicine from 98 to 138 C.E., denied that the womb was an animal, but he still acknowledged the existence of a condition that was diagnosed as hysterical suffocation. Soranus's idea that the womb could cause choking was widely shared among many ancient experts, even though they disagreed about whether the womb could actually move around in the body and whether it was literally a wild animal.[28]

In tracing these physical explanations of hysteria throughout the centuries and across ancient texts from multiple cultures, we see much variation and conflict, but we also see some ideas that persistently recur. In particular, experts who espoused this explanation of hysteria throughout these centuries characterized the womb as having the ability to move throughout women's bodies, suggesting that it causes health problems by pressing on other organs. In some of the ancient texts, the womb's movement is depicted as being like that of a wild animal in search of food; in other texts, it is depicted in a way that is more similar to our modern understanding of the ways in which an organ behaves, such as with the idea that the womb causes suffocation by rising and pressing on the lungs and other organs. It is amid this confusion and these competing explanations that the humoral explanation, which is the focus of the next section, emerged.

Humoral Explanation of the Womb

The humoral explanation of hysteria emerged alongside—and was sometimes intermingled with—the physical explanation. Beginning in the Ancient Greek texts, experts grappled with uncertainty about whether the womb could actually move around inside women's bodies, as had been suggested in some texts. The humoral explanation of hysteria, which is usually attributed to Galen in the second century C.E.,[29] was an alternative explanation for the womb's role in hysteria that emerged in response to this uncertainty. Instead of accepting the idea that the womb literally moved within the body, Galen suggested that hysteria was caused by abnormal retention of uterine secretions, offering the basis for a humoral explanation.

Despite Galen's questioning, the idea that the womb actually moved inside the body persisted for several centuries; without access to modern understandings of anatomy, these remained largely theoretical issues that could only be addressed through speculation. As anatomical understanding became more sophisticated, however, the humoral explanation for hysteria started to assume more nuanced forms. As early as the thirteenth century, some experts started looking for anatomical evidence to refute definitively the idea that the womb actually moved. For instance, a scientist named Mondino de Luzzi provided anatomical evidence to verify that the womb existed in a fixed location inside a woman's body.[30] In light of the new types of anatomical evidence that were becoming available at this time, the womb started to assume a different shape— perhaps in a way that resembles today's understanding of an organ, rather than a wild animal or a mix of fluids. However, an important aspect that connected this new shape to earlier versions of the womb concept was that the womb—and concepts about it—still provided a basis for explaining hysteria. However, experts had to find different ways to account for the womb's role in hysteria in the new explanations of hysteria. One explanation was Mondino's suggestion that the stationary womb could cause hysteria when it "affects the heart or lungs by vapours driven upwards."[31]

As early as the sixteenth century, in the writings of physicians such as Andreas Vesalius and Ambroise Paré and philosopher René Descartes, the concept of hysteria was formulated in ways that correspond increasingly with modern scientific understandings of the human body. These experts followed Mondino's earlier lead by developing new explanations of hysteria that still characterized it as an affliction of the uterus but without assuming that the uterus could have literally wandered around inside women's bodies.[32] Although these experts offered many different theories to explain how the uterus caused hysteria, they widely agreed that the uterus, in some manner, was the source of the affliction. According to their arguments, their understanding of the womb gradually transformed into something more like our current understanding of what an organ is; thus, they tended to move away from Hippocrates's idea that the uterus could literally wander around in the body and toward Galen's humoral understanding of the womb's role in hysteria. From the thirteenth century to the seventeenth century, the womb was transformed from being thought of as a vaguely defined entity (whether animal-like or mechanistic) that could only be imagined and into one that is more like our current understanding of an organ—that is, as having a specific function and relationship with other organs inside the body. The key distinguishing feature of this revised understanding, which can be traced at least as far as Galen's humoral explanation, was that the relationship between the uterus and the brain was sustained through some kind of communication network inside the body that allowed for organs to interact without making direct contact with each other.[33]

The increased recognition of the brain's role in the body led to some additional transformations in the humoral explanation of hysteria. For instance, at the beginning of the seventeenth century, a physician named Edward Jorden started to offer a sympathetic explanation—namely, that the womb had sympathetic connections with other organs, including the brain, and that is how it caused hysteria without directly pressing itself on these other organs. Jorden came to the defense of some hysterical women who were being accused of witchcraft by using Galen's humoral understanding to refute supernatural explanations that affiliated hysteria with witchcraft.[34]

Although these explanations, over the centuries, posited an increasingly complex relationship between the womb and other organs in the body, they continued to repeat many key ideas from the ancient texts, such as the notion that hysteria could be described as the feeling of a large globe swelling and rising in the abdomen. In fact, through the sixteenth century, physicians who treated hysteria still relied largely on the ancient texts of physicians such as Hippocrates and Galen.[35] As I have demonstrated, however, they were transforming and twisting the ideas from these texts, and as a result, the womb itself was being reshaped as more of an organ, in the modern sense, and less like a figment of the imagination. Nonetheless, the basic understanding of the womb, and its role in causing hysteria, continued to be based on the authority of the ancients, with new empirical observations always being made to fit into this ancient framework.

In the seventeenth century, the uterus gradually morphed into something that could be used to support more precise arguments about the means of diagnosing hysteria and the kind of evidence that should inform a diagnosis. Seventeenth-century physician Thomas Sydenham was one of the first to claim that he could offer a clinically based understanding of hysteria. Unlike many of his predecessors, Sydenham claimed that his philosophy of medicine was based in practice. He followed Bacon in emphasizing the importance of practice and observation over theoretical speculation. Of course, the ancient physicians had also claimed that their explanations were based on patient observations, but Sydenham followed Bacon in claiming to reverse the order so that empirical observation came first, then speculation. This can be seen, of course, as part of the larger trend in this period of rejecting the authority of ancient texts that had dominated perceptions of truth for so many centuries and a reflection of thinkers such as Bacon who believed that "sciences based on crafts start at their least and improve with time, while the converse is true of those that start with speculation."[36] For at least a thousand years prior to Sydenham's writings, medical understandings of "hysterical affection" had begun with theoretical explanations and were followed by clinical observations, so Sydenham's claim to reverse this order was significant. Along these lines, Sydenham was known for his detailed, written descriptions of clinical observations of actual patients. Thus, in rhetorical terms, the seventeenth-century womb came to be reshaped yet again as a result of the fact that clinical evidence was gradually beginning to replace theoretical speculation as the basis for medical diagnoses.

Despite these variations in experts' understandings of exactly how the uterus caused hysteria, the idea that the uterus was the primary cause of hysteria seems to have been universally accepted until the seventeenth century, when Thomas Willis (1621–1675), an English practitioner, offered one of the first nonuterus-based explanations of hysteria, suggesting that hysteria was caused by animal spirits acting on the brain.[37] Following Willis's lead, the separation between the uterus and the brain became more pronounced in the seventeenth century, and the brain of the hysterical woman received increasing amounts of attention.[38]

By the end of the seventeenth century, the womb was assuming a new shape that positioned it not so much as a distinct organ, separate from the brain, but as an organ that existed in a complex relationship among the brain and other parts of the female body. This complicated shape became increasingly prominent throughout the seventeenth, eighteenth, and nineteenth centuries as older shapes of the womb—resembling those presented in ancient texts—started to mix with more modern explanations. The result was a complex and often contradictory set of ideas about the brain-uterus relationship in the female body that was still circulating in the authoritative texts of the nineteenth century. This line of thinking perpetuated many curious ideas about anatomical connections between the brain and central nervous system and the uterus. There were nineteenth-century physicians who claimed, for example, that they had observed such phenomena as a "connection of the uterine nerves with those of the larynx" to have been a frequent cause of vocal problems in women and that "local [pelvic] irritations acting upon the central organ [brain]" could have been the cause of a conditioned called "lactational insanity."[39]

Moral Explanation of the Womb

The moral explanation can also be seen as having its roots in the ancient texts, insofar as these texts often recommended the socially sanctioned feminine acts of marriage, intercourse, and pregnancy as the best cures for hysteria. Over the centuries, these moral ideas have taken on various shapes, sometimes pertaining to social practices such as marriage and reproduction, and sometimes pertaining to supernatural elements believed to have been factors that caused hysteria. Elements of the supernatural have been present, alongside other explanations, in recommended treatments for hysteria since the ancient Egyptian papyri.[40] But the moral dimension of hysteria became more pronounced in the medieval period as hysteria came to be affiliated with witchcraft. Many still believed in the wandering womb as the source of this condition, but religious incantations were sometimes used, instead of physical treatments like odors, to restore the womb to its proper position. Although improved scientific understandings of the physiologies of the brain and the uterus eventually caused the witchcraft explanation to subside, the moral dimension of hysteria persisted for several centuries, evolving to assume different shapes that correspond with the social influences of subsequent eras.[41]

For example, by the seventeenth century, fears about a supernatural element of hysteria had been transformed into concerns about women's behaviors as a cause of hysteria. Historian Jeffrey M. N. Boss provides a composite definition of hysteria as it was described in printed books available at the beginning of the seventeenth century. Although there were subtle differences among the experts, Boss suggests, it was widely accepted that hysteria "comes on in fits in a woman who is, characteristically, in good health otherwise." Boss's composite definition suggests not only that experts in this era continued to emphasize the moral aspects of treating hysteria, but also that these moral concerns were transforming from concerns about witchcraft or demons into concerns about the hysterical woman's social life. As he says, the hysterical woman in this era was "commonly a maiden, widow or spinster . . . or a woman who is failing to menstruate."[42] These fears about women's social behaviors intensified in subsequent centuries, and by the nineteenth century, they had transformed into concerns about social influences that were causing women to do more intellectual work than they had done in previous eras, thereby endangering their bodies (the uterus in particular and, hence, women's reproductive capacities).[43]

Nineteenth-Century Hysteria: Conflicting Explanations

Aristotle's Masterpiece and Hollick's collection of texts, which provide the focus for this section's analysis, seem to have been accepted by nineteenth-century consumers as authoritative sources of information about a wide array of health- and sex-related topics. However, when it came to hysteria, these texts cited many different popular and scientific ideas from the long history of this primarily female health condition, and the authors even cited some ideas that had been refuted and disproven by previous generations of experts. As we will see, the authors of these texts incorporate all three explanations of hysteria—physical, humoral, and moral—sometimes even juxtaposing these explanations within the same chapter or paragraph.

Observing how the authors treat these outmoded beliefs is in itself an interesting rhetorical phenomenon: at times, the authors seem to accept the older ideas, but at other times, they make an explicit effort to dissociate from these older ideas. When they dissociate themselves from the older beliefs, they either suggest that their own evidence is scientifically superior to that which was offered in the ancient texts, or they use moral reasoning that suggests that their own beliefs in Christianity provide them with a moral framework that is superior to that which was available to the authors of the ancient texts. Thus, there are many places in *Aristotle's Masterpiece* where the author distinguishes between the ancients and the moderns and aligns himself with the moderns. This tendency can result in a quite confusing presentation of the author's beliefs. For instance, in a section titled "Of the Womb in General," the author refers to an ancient belief that suggested that males are conceived on the right side of the womb because its proximity to the liver causes it to be warmer, whereas females are conceived on the left side of the womb because its proximity to the spleen causes it to be colder. After acknowledging that this belief originates in the ancient texts of Hippocrates, the author then distinguishes his own modern understanding of the belief from that of the ancients, suggesting that Hippocrates allowed for some exceptions to this rule, but the moderns believe it to be always true. Specifically, the author says that "most of our moderns hold the above as an infallible truth, yet Hippocrates holds it but in general."[44] He then quotes Hippocrates at length and implies that Hippocrates did not have access to entirely accurate information.

Aristotle's Masterpiece also includes an interesting mix of ancient beliefs, such as humoral theories, and what would have been considered modern Christian beliefs at the time. This is seen, for instance, when the author discusses procreation, which he says is an "injunction" that was "imposed" on men and women "by God at their first creation, and then again after the deluge." He then talks about the necessity of both men and women, and the complementarity of their different natures, suggesting that propagation would not be possible without these complementary qualities, but in so doing, he reverts to humoral theories of the body: "Man therefore is hot and dry, woman cold and moist; he is the agent, she the patient or weaker vessel, that she should be subject to the office of the man."[45] The result of this tendency to alternate between associating and dissociating from these ancient beliefs is that it creates confusions in the texts that I analyze below, especially for a reader who tries to make sense of the contradictory ideas about hysteria that circulated in the nineteenth century. Nonetheless, it is possible to trace elements of all three explanations of hysteria—physical, humoral, and moral—as they persisted in these authoritative nineteenth-century texts.

The physical explanation of hysteria occurs in several places throughout both of these texts. For instance, in chapter VI of *Aristotle's Masterpiece*, "The Suffocation of the Mother," the author echoes Soranus's description of a condition in which the womb can rise in the body, causing it to press on the respiratory organs and also, indirectly, on the brain. The author also refers to the condition as "uterine suffocation" and repeats a common theme found in many early texts about hysteria: the notion that the hysterical woman can sometimes be mistaken for dead. He says regarding one patient that her friends "sent for a surgeon to have her dissected," but then the woman "began to move, and with great clamour returned to herself again." The author then describes more precisely what he understands to be the role of the womb in hysteria: "The part affected is the womb, of which there is a twofold motion—natural and symptomatical. The natural motion is, when the womb attracteth the human seed, or excludeth the infant or secundine. The symptomatical motion, of which we are to speak, is a convulsive drawing up of the womb." He then describes several possible causes of this "symptomatical motion": "The cause usually is in the retention of the seed or the supression of the menses, causing a replention of the corrupt humours in the womb, from whence proceeds a flatuous refrigeration, causing a convulsion of the ligaments of the womb. And as it may come from humidity or repletion, being a convulsion, it may be caused by emptiness or dryness. And lastly, by abortion, or difficult childbirth."[46] The author of *Aristotle's Masterpiece* then distinguishes hysteria from a number of other diseases that it was often confused with, including apoplexy, epilepsy, syncope, and lethargy.

Hollick's texts include similar ideas about how the movement of the womb caused hysteria. In *The Diseases of Woman*, the chapter about hysteria describes this condition as the "most mysterious, confusing, and rebellious of all female diseases." Hollick traces the origins of hysteria to ancient Greece and repeats the long-accepted idea that hysteria was a "uterine affection" but then says that "the symptoms of this disease comprise . . . those of nearly every other disease under the sun." He starts with the "hysterical fit," which was manifested through symptoms that ranged from "palpitations" of the heart to uncontrollable laughter or crying. He proceeds to describe a truly horrible collection of symptoms that included phsyical symptoms such as vomiting and many other symptoms that we would characterize today as phsyical manifestations of anxiety. But movement of the uterus is clearly present in his description as an underlying cause of these symptoms.[47] Echoing the idea that had become a refrain in the ancient texts as well as *Aristotle's Masterpiece*—namely, that the rising uterus could actually have suffocated the mother—Hollick says that when the hysterical fit began, the woman would have "[felt] in some part of the abdomen a sensation as if a large round ball, or globe, was moving about."[48]

Hollick's other text, *The Origin of Life and Process of Reproduction in Plants and Animals*, also includes a section about displacement of female organs. In this section, he begins with the womb and expresses many ideas that echo directly from Hippocrates and the ancient Greeks. He says that "the different female organs are liable, from a variety of causes, to be displaced," but "the womb is most frequently found outside of its proper situation."[49] He then provides information about several different types of womb displacements that could occur in a woman's body. These include, for instance, "prolapsus uteri, or falling of the womb" (figure 3) and "anteversion and retroversion of the womb" (figure 4). Although he echoes ancient ideas about the displacement of the womb, his detailed diagrams and explanations offer a nineteenth-century twist on these ideas. He also echoes an idea that was coming to be commonly accepted at this time—namely, that the uterus was connected to the central nervous system in a much more precise way than had previously been described. He identifies three

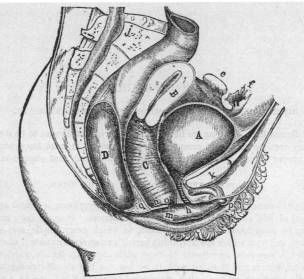

FIGURE 125.—*Ltaeral Section of the Female Pelvis, to show the position of the organs in their natural state.*

A. The bladder. *B.* The womb. *C.* The vagina. *D.* The rectum. *e.* The right ovary. *f.* The right Fallopian tube. *g.* The os tincæ, or mouth of the womb. *h.* The meatus urinarius, or mouth of the bladder. *i i.* The small intestines. *j j.* The back bone. *k.* The pubic or front bone. *l.* The right external lip, or labium. *m.* The right internal lip, or nymphæ. *n.* The hymen. *o.* The opening through the hymen. *q.* The clitoris. *p.* The perineum.

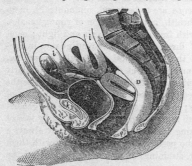

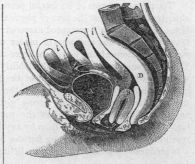

FIGURE 126.—*Lateral Section of the female Pelvis, to show the position of the Womb in the First Stage of Prolapsus.*

A. The bladder. *B.* The womb, which is fallen down nearly to the middle of the vagina, which is much enlarged, owing to the womb being forced down into it. *D.* The rectum, also much compressed. *i-i.* The small intestines, also fallen down after the womb.

FIGURE 127.—*Lateral Section of the Pelvis, to show the position of the Womb, and other Organs, in the Second Stage of Prolapsus.*

A. The bladder. *B.* The womb, now fallen to the bottom of the vagina, which is much enlarged, and nearly filled up by the fallen womb. *D.* The rectum, which, like the bladder, is severely pressed. *i.* The small intestines, still following the womb.

3 | Images showing the normal placement of the female organs and two stages of the prolapsed womb. From Frederick Hollick's *The Diseases of Woman: Their Causes and Cure Familiarly Explained* (New York: Burgess, Stringer, 1847).

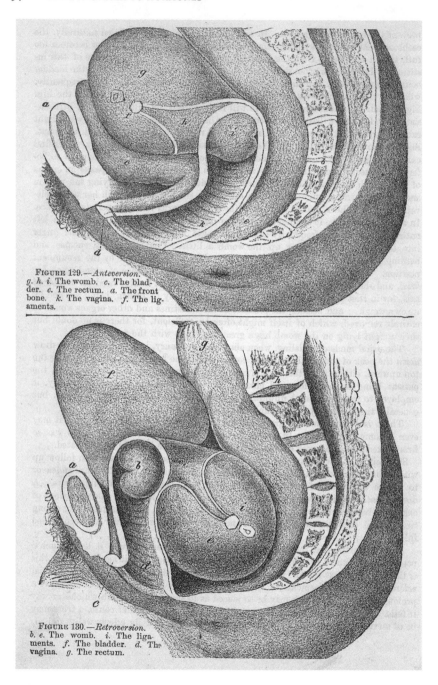

FIGURE 129.—*Anteversion.*
g. h. i. The womb. *c.* The bladder. *e.* The rectum. *a.* The front
bone. *k.* The vagina. *f.* The ligaments.

FIGURE 130.—*Retroversion.*
b. e. The womb. *i.* The ligaments. *f.* The bladder. *d.* The
vagina. *g.* The rectum.

4 | Images showing the anteversion of the female organs and the retroversion of the womb. From
Frederick Hollick's *The Diseases of Woman: Their Causes and Cure Familiarly Explained* (New York:
Burgess, Stringer, 1847).

stages of the "fallen womb" disease: the first, in which the womb has "merely *settled down* a little,"[50] and then the two subsequent stages in which it has fallen further. He discusses extensively several possible cures, including insertion of pessaries (of different varieties), and the most extreme remedy, which is surgical removal of the uterus.

In addition to the physical explanations that describe the womb as pushing on other organs, both of these nineteenth-century texts include other passages that depict the influence of the womb in a way that resonates more with the humoral explanation. In some places in these nineteenth-century texts, we even see ideas that sound similar to what we currently understand as the nervous system or the endocrine system. In these passages, the womb is depicted not so much as having directly pressed itself on other organs, but rather as having relied on some vaguely defined messaging system to communicate with other organs throughout the body. In some of these descriptions of the womb's role in hysteria, we start to see how these nineteenth-century authors distance their own beliefs from those of the ancients, even though these authors still give some credence to the ancient beliefs. For instance, when the author of *Aristotle's Masterpiece* refers to the ancient idea of the wandering womb, he sometimes describes communication between the womb and other parts of the body in language that sounds as if these ideas might be scientifically verifiable. This language can be understood as an attempt to meld the ancient idea of a wandering womb with ideas about the brain and central nervous system that had started to crystallize in the seventeenth century. As the author says, "with the evil quality of the womb, the whole body stands charged, but especially the heart, the liver, and the brain." He then describes how the womb "communicates" with each of these organs: it communicated with the heart through arteries, with the liver through veins, and with the brain through "the nerves and membrane of the back."[51] Book II of *Aristotle's Masterpiece*, which is titled "A Private Looking-Glass for the Female Sex," includes some similar ideas in a section titled "Treating of several maladies incident to the womb, with proper remedies for the cure of each." In describing the womb as an organ, the author says that it included "two angles," with different surfaces or compositions: "the outward part of it nervous and full of sinews, which are the cause of its motion, but inwardly it is fleshy."[52] Such language is interesting because it attempts to describe a specific nervous dimension to the womb. Although the information is obviously inaccurate, according to today's knowledge, it can be seen as an early attempt to meld ancient ideas about the wandering womb with the scientific knowledge of the body that was rapidly emerging in several new subdisciplines of medicine in the nineteenth century.

In Hollick's texts, we also see an interesting mix of ancient ideas that treat the womb's literal displacement as the cause of hysteria, and modern beliefs that posit a more complex and abstract relationship between the womb and the rest of the body. The following passage in Hollick's text is an important indication of one way in which the relationship between uterus and brain was understood at the time:

> The *nerves* of the uterus are connected with those of almost every other organ in the body, as well as with the spinal marrow, and this explains why it has such extensive and complicated sympathies. There is, in fact, scarcely any organ in the body but what may suffer, and appear to be diseased, though perfectly healthy, merely from disease of the womb, which may nevertheless give but few or no indications of disease itself. . . . Palpitation of the heart, bilious derangements, considerable spinal irritation, inflammation of the bowels, difficulty of breathing, and dyspepsia, besides a host of minor derangements, are often produced by falling of the womb. When this is the case we must, of course, look for a cure only by restoring that organ to its place; but how could a female feel satisfied that any treatment of the womb would relieve the heart, or stomach, unless she knew how intimately it was connected with them?[53]

It is interesting that in this passage Hollick mentions physical displacement of the womb as one possible cause of hysterical symptoms, but he also emphasizes a humoral explanation when he says that the womb "has complicated sympathies" with many other organs in the body. One distinguishing feature of Hollick's nineteenth-century text is that there is some effort to be scientific. As he outlines the possible ways in which the womb can be displaced, he bases his claims on observations that have been made of the uterus at different times. Many of his observations pertain to childbirth, and he speaks of such interventions as a physician reaching in and physically manipulating the womb back into place. So there is a more tangible, concrete element to the displaced womb than was apparent before, and there is also this distinct emphasis on the relationship to the brain and central nervous system, but this latter notion is based purely on speculation, without any direct scientific evidence.

In sum, then, what is interesting in this book (at least in this extensive section on displacements of the womb) is the extremely detailed focus on the physiological characteristics of the uterus and the myriad pathological conditions that could arise from its displacement. It is important to note that these different explanations coexist in Hollick's texts, sometimes even within the same chapter or paragraph. For instance, although he suggests in the passage quoted above that the womb is connected to the central nervous system as a way to explain how it might have a variety of other effects throughout the body, this particular section focuses just on the womb itself and surrounding organs such as the bladder. Later in the book, when he is discussing "engorgement of the womb," he mentions the possibility that the patient could have become "hysterical" as a result of this condition.[54] This is an interesting blurring of the boundary between uterus and brain, and later when he discusses cures, this becomes even more apparent. As he says, "sometimes it is necessary to advise a state of *single-ness* for a time, but at other times marriage will be beneficial."[55] Then in the next chapter, titled "Derangement of the Functions of the Female Organs, and of the Nerves," he focuses more directly on the womb's relationship to the nerves. In the first two sections of the chapter, he discusses emotional factors (stressful events, and emotions such as "anger, jealousy, or excessive joy") that can cause menstruation to be delayed.[56]

The section that is specifically devoted to hysteria, by contrast, is a later subsection within this same chapter, and it suggests a much more complicated and speculative relationship between the womb and the rest of the body. This is interesting because, once again, it posits the uterus as not just a stand-alone organ whose sole purpose was reproduction but as fundamentally connected to the nervous system—that is, it seems to portray the womb's displacement much less literally and surrounded by more mysticism than does the other chapter that addresses literal displacement of the uterus. Rather than describe specific physical attributes of the uterus, or displacements within the female body, in the section on hysteria Hollick focuses on mysterious behaviors that seemed to connect the uterus to other parts of the body. He presents these behaviors as if he had actually observed them, but they are clearly based on speculation rather than any scientific evidence. For instance, he says that after a patient recovered from a hysterical fit, "an abundant mucus secretion flow[ed] from the vagina, through previously it was unusually dry and constricted." Then he attributes this symptom to the uterus, without any direct evidence: "This is a proof how much the uterus sympathizes in this disease."[57]

In *Artistotle's Masterpiece*, we see relatively little attention to the moral explanation of hysteria. Throughout the text, ideas from the Bible are often juxtaposed with ideas from other ancient texts. For example, the title of chapter VI is "Of the happy state of Matrimony, as it is appointed by God; the true felicity that redounds thereby to either Sex; and to what end it is ordained." In the first paragraph of this chapter, the author refers to the story of Adam and Eve as a justification for the natural happiness that ensues from marriage: "And truly a wife, if rightly considered, as our grandfather Adam well observed, is or ought to be esteemed of every honest man, 'bone of his bone, and flesh of his flesh,' &c."[58] Right after this passage, the author cites a similar idea from Xenophon's *The Economist*, an ancient Greek text that is dated around 400 B.C.E.: "Wherefore, since marriage is the most safe, sure, and delightful situation of mankind, who is exceeding prone, by the dictates of nature, to propagate his like, he does in no ways provide amiss for his own tranquility who enters it, especially when he comes to maturity of years."[59] Although these general ideas about the benefits of marriage are repeated frequently throughout the text, and these ideas certainly have moral overtones, there is relatively little attention to the moral dimensions of the causes and treatments of hysteria in *Aristotle's Masterpiece*.

When moral dimensions of hysteria appear, they are closely tied to the physical dimensions of the disease. For example, the author mentions two possible causes of hysteria, both of which are treated as physical causes: "suppression of the menses" and "retention of the seed." Regarding the latter, he suggests that "a good husband will administer a cure."[60] Other than this piece of advice, which could be interpreted as having moral overtones, the cures that are offered in *Aristotle's Masterpiece* involved strange mixtures of various substances, some of which are attributed to Galen and most of which reflect a humoral understanding of the body.

By contrast, moral explanations for hysteria occur frequently throughout Hollick's texts. These explanations assume several shapes. One version of the moral explanation is that women can actually work themselves into a hysteria-like condition through failure to control their emotions. As he says, "many females can work themselves into hysterics . . . particuarly when angered, slighted, or disappointed."[61] Nonetheless, he then proceeds to proclaim that the uterus was the root cause of hysteria. This is where we start to see the peculiar set of beliefs that governed nineteenth-century understanding of the uterus-brain relationship. The passage below effectively captures the confusing set of beliefs that governed mid-nineteenth-century understandings of the uterus and brain, and is another example of these nineteenth-century authors' citing ancient beliefs while simultaneously dissociating themselves from complete agreement with the ancients:

> In regard to the starting point, or original seat of Hysteria, there seems to be no doubt of its being in the Uterus, which becomes subject to a peculiar excitement, or disturbance, that exerts a wonderful sympathetic influence on the whole system. The Uterus, it must be remembered, is the *controlling* organ in the female body, being the most excitable of all, and so intimately connected, by the ramifications of its numerous nerves, with every other part. The multitudinous and diversified symptoms attending its derangements need not therefore surprise us, nor need we wonder that they are not found in other diseases. The ancients compared the Womb in the female body to *another living being,* controlling and directing the body in which it existed! It should also be remembered, in relation to hysteria, that it is most frequent at that age, and in those temperaments in which the uterine system is most active.[62]

Hollick then says that sometimes "the nervous excitement and vascular turgescence of the uterine organs determine the character of the mental disorder," and he speaks of "local or uterine irritation" as another possible factor.[63]

In other places in his texts, Hollick seems to adopt very different ideas, attributing hysteria to external social factors (such as city living) rather than just the womb itself. In offering suggestions to prevent hysteria, he again reverts to many social and moral factors, suggesting that "an indolent, a luxurious, and an unoccupied life" could lead to, among other consequences, "irritation and turgescence of the generative organs." For this reason, one of the most important mechanisms of prevention that he recommends is "moral and physical education of females about the period of puberty in preventing hysteria." He then says that better education in this regard would lead to "a race of females possessed of stronger minds, and better able to make good wives and healthy mothers."[64]

Interestingly, Hollick says that the causes of hysteria are "as abscure as the symptoms are diversified." The possible causes that he lists seem to be mostly external factors: "weak constitution, scrofula, indolence, a city life, bad physical and moral education, nervous or sanguine temperaments, the over-excitement of certain feelings, and religious or other enthusiasm."[65] Some of the causes of hysteria that he hints at will come to be understood as hormonal soon after this historical period. For instance, he says that hysterical fits were "most common between puberty and the change of life" and are frequently experienced by "those in whom the change of life is about to take place." He also suggests that individual personality traits made some women more susceptible to hysteria, such as being "capricious in their character" or "exceedingly excitable."[66] In this section of the book, Hollick makes one mention of the uterus as a possible cause. Even here, however, he treats the uterine disease as a separate condition, simply suggesting that this condition might have made the woman more susceptible to hysteria, but he still leaves room for the moral dimensions that he emphasizes elsewhere in this section of the book: "Various other diseases, particularly of the uterine organs, may also produce hysteria."[67] This differs from the belief articulated in earlier centuries—namely, that physical displacement of the womb was the actual cause of hysteria. Instead, in this section of this book, Hollick tends to emphasize social conditions and habits of modern women as the primary factors that would lead the susceptible woman to become hysterical. The womb remains implicated in this understanding of hysteria, but it is subject more to social intervention than to medical intervention. Of course, this advice echoes the ancient advice of intercourse and pregnancy as a cure for hysteria, but Hollick's advice is more nuanced, as it is adapted to correspond with the social milieu of the nineteenth century.

The language in Hollick's book, first published in 1847, is a perfect illustration of the confusing, conflicting sets of beliefs that circulated in the nineteenth century around women's reproductive and nervous systems. In an era that demanded increasingly scientific explanations, as medicine was struggling to tie itself more to science than to myth, there was clearly a need for scientific explanations of the brain-uterus relationship, yet Hollick, as an expert, ultimately resorted to ancient beliefs, such as when he mentions that the ancients believed the womb was a living creature. From a rhetorical perspective, it is interesting to note how even as he mentions these ancient beliefs, he dissociates his own position from them. Although he does not refute the ancient beliefs, he seems to suggest that his version of the truth is more credible. He does not literally believe the uterus was a living creature, as the ancients did, but he believes in some of the effects that led the ancients to characterize the uterus in this manner. As a means of further distinguishing his own beliefs from those of the ancients, he also adds the moral dimension, promoting better sex education at the time of puberty, for instance, as a means of preventing hysteria.

Conclusion

It is hard to think of another health condition that has caused as much confusion as hysteria, or one that has been subject to such a wide variety of expert interventions and treatment recommendations. The idea of the wandering womb is clearly an old one, and the conflicting explanations of the womb's movement that we see in nineteenth-century popular texts can be traced to multiple conflicting explanations that first appeared in the most ancient texts that are available to us. This chapter's analysis has depicted these explanations as different shapes that the womb has assumed, highlighting how each generation of experts found ways to reshape this female organ to meet their own needs, causing the womb itself to possess topological characteristics in that it was able to sustain for many centuries repeated acts of twisting, bending, and reshaping while still preserving its central place in expert understandings of women and their female problems.

In the Western medical tradition, Hippocrates is usually cited as the first physician to document cases in which the wandering womb led to symptoms of hysteria. But in light of the evidence recorded in Egyptian and Chinese texts, it is apparent that, when Greek experts started documenting their versions of these beliefs around the fifth century B.C.E., they were not beginning from a blank slate; rather, they were documenting and offering further speculations about a set of beliefs that had already been circulating elsewhere. Although the basic ideas about the womb as having pressed itself on other organs to cause hysteria sound similar in the texts that this chapter has analyzed, the womb's role in hysteria becomes much more confusing and contradictory when we consider several other ideas that are expressed elsewhere in these texts. In these other places, the womb's impact on the body is explained in a way that depends on the newly emerging understanding of how different organs could communicate with each other through mechanisms that would soon be understood as the central nervous system and the endocrine system. As I have demonstrated in this chapter, even nineteenth-century texts preserved some fundamental ideas, such as the wandering womb, from ancient times. However, the authors further manipulated and reshaped the womb to meet the needs of their contemporary audiences. One of the most puzzling aspects of the rhetorical situation of these nineteenth-century texts is that these texts preserved ideas that were solidly refuted centuries earlier. This is when hysteria becomes even more confusing than other medical conditions that are often the object of research in the history of medicine. Although the ancient texts of medical experts such as Hippocrates, Paracelsus, and Galen retained their authoritative status for several centuries after they were written, historians of medicine have provided convincing evidence to illustrate that the womb did not maintain a constant shape as these ancient texts were cited, discussed, and incorporated into medical practice throughout these centuries.

As we can see by the popularity of the best-selling *Aristotle's Masterpiece* and of Hollick's texts, ancient ideas such as the womb as a central influence on women's health and behavior were still taken quite seriously by the nineteenth-century reading public, both in Europe and the United States. Although, as we will see in the next chapter, some nineteenth-century experts were working hard in newly emerging subspecialties of medicine to provide new explanations, the ancient beliefs still held popular sway. In short, it is clear that the idea of the uterus as a cause of hysteria persisted well beyond the ancient texts in which it originated, even as experts' explanations for the exact manner in which the uterus caused the disease continued to be a subject of contention.

The different ways in which the womb has been reshaped over the centuries to support different explanations of hysteria can be seen as a reflection of Serres's notion of time as topological. Along these lines, Serres's notion of time as topological has a distinctly gendered dimension that makes this concept especially appropriate as a framework for understanding a health condition that has been seen as primarily one afflicting females. In Serres's words, the fact that we have overlooked topological dimensions of time and instead favored linear understandings has everything to do with the fact that "we have never known what a tissue is, never noticed or listened to women."[68] This also relates to the concept of adelo-knowledge, as explained by Steven Connor: "Serres describes [adelo-knowledge] as the opposite of analysis, or the separating of things one from another (for topological transformation disallows cutting)."[69] This description of adelo-knowledge nicely conveys this chapter's approach to the ever-changing womb in relation to the other components that have been affiliated with hysteria over the centuries. Rather than dividing expert understandings of the womb into distinct eras, the effort in this chapter has been to notice what is preserved in the basic substance of this female organ even as it has been endlessly twisted and reshaped to meet the needs of each new group of experts that have emerged in the history that extends from ancient texts to the nineteenth century.

In subsequent chapters, we will see how these conflicts and contradictions created an ideal source from which a concept such as hormones could emerge and seem to offer some clarity and light about the previously mysterious and dark subject of women's bodies. Although rhetorical and historical scholars have widely understood the nineteenth century as the distinct moment when Western medicine became more closely aligned with science, this chapter and the next both expose—through the example of hysteria—that the nineteenth century was actually a time of intense conflict, noise, contradiction, and disagreement between ancient and modern beliefs. These differing perspectives appear among the texts of numerous experts who were writing about hysteria at the time, often within the same texts. We will see in later chapters that, from a historical perspective, the recalcitrance of hysteria and the resulting frustration of experts who could not figure it out were key components of the historical context that preceded the 1905 emergence of the term *hormone*.

As I will continue to emphasize in subsequent chapters, even the authors who tried to provide the most scientific explanations for female problems still reverted to ancient ideas, some in more explicit ways and some in more implicit ways. Among the latter, for instance, were scientists who focused on the brain rather than on the uterus; their explanation of the female brain, or female intellect, often made the female brain resemble the mysterious womb from days gone by. Although they greatly desired to break away from the ancient ways of thinking and to unite medicine with science instead, nineteenth-century scientists and physicians were never quite able to achieve this union to the extent that they desired. They lacked adequate evidence and explanations, which is one reason why they occasionally resorted to ancient beliefs. We might speculate that they did so because the general public still very much accepted these ancient beliefs, as demonstrated by the continued use of smelling salts. Invoking these ancient ideas helped their audiences accept their "modern" ideas.

In these ways, this chapter's analysis of hysteria as a rhetorical phenomenon has enacted the kind of adelo-knowledge that Serres hints at. As a result, the analysis does not necessarily clarify anything; rather, the aim has been to foreground the confusing mix of contradictory ideas that were still circulating in the nineteenth century, even as science and medicine were making a concerted effort to be more systematic and more closely tied to each other. In so doing, this chapter's analysis has also noted the points where rhetorical theorists' understanding of concepts such as invention reach their limits when it comes to accounting for scientific arguments about a subject such as hysteria that is so much more explicitly enmeshed in cultural and scientific understandings of the sexed body than are the scientific objects that rhetorical theorists have typically used in developing concepts such as invention. I have already noted how hysteria's historical trajectory differs from that of most other diseases that medical historians have studied, but it is also important to note how it differs from the typical subjects of rhetoric-of-science scholarship. Such scholarship has tended to cluster around certain key scientific controversies or breakthroughs, including those that stemmed from the work of Darwin,[70] the discovery of DNA,[71] and current and past discourses of genetic science and eugenics.[72] Some of this scholarship is certainly beginning to grapple with the material realities of bodies and the norms of sex and race that govern these realities,[73] and we see additional examples along these lines in recent rhetoric-of-medicine scholarship.[74] But to an even greater degree than any of these other objects of study, hysteria forces us to confront the limits of a scholarly approach that begins from the assumption that science progresses in a manner that can, at least in retrospect, be characterized as proceeding through historical eras that are defined by expert arguments that either win or lose—that is, these arguments either take or cede control of a given moment in this long progression. Discovery and invention are forms of movement that have received much attention from scholars in history and rhetoric of medicine. This chapter has continued that focus but has troubled these previous approaches because hysteria has been such a difficult case: medical experts have been defied in their attempts to pin it down, and historians of medicine have been prevented from identifying a single moment as a big breakthrough or discovery regarding the condition, if we even wanted to do so.

In short, in the case of hysteria—even more so than for other medical conditions that have long histories—it is especially appropriate to invoke Serres's metaphor of the Seine River and to draw more attention to the multiple conflicting undercurrents that underlie the river's seemingly smooth surface than to the smooth surface itself. With its ability to twist, morph, shape-shift, and move unpredictably around inside a woman's body, the womb as depicted in the earliest recorded texts available to us provides the ideal subject matter for glimpsing this phenomenon that, as we will see in subsequent chapters, continued well past the dissolution of hysteria as a legitimate disease category.

3

Charcot's Circus | Nineteenth-Century Science of Hysteria as a Moment of Stasis

For many years I walked through my wards like a blind man, never seeing hysteria, not because it was not there, after all it is common, but because I did not know how to look at things.

—Jean-Martin Charcot, *Lectures on the Diseases of the Nervous System*

Stasis, which translates literally from Greek as "stopping place," has been used in classical rhetoric to understand the different kinds of stopping places, or forms of disagreement, from which arguments can unfold. This way of thinking about stasis has offered a fruitful way to illuminate how and why some arguments get stuck at a certain point beyond which some or all of the participants in a dispute are unable to move. For instance, Christa Teston and S. Scott Graham reveal how a failure of stakeholders to agree about what counts as a clinical benefit and what counts as legitimate evidence prevented any meaningful form of public participation in a recent FDA hearing about the safety and efficacy of the use of Avastin as a breast cancer treatment drug.[1] Their analysis invokes the concepts of evidentiary and methodological stases to explain why even though public participants whose expertise derived from successful treatment that included the use of Avastin were invited to participate in the hearing, their testimonies were not perceived as legitimate sources of knowledge, so the hearings did not result in any new knowledge production, and the FDA's stance on Avastin was not changed.

Other analyses offer different interpretations of this concept, using stasis theory to illuminate how change can occur in science and medicine. In these studies, we see that even if the moments of stasis—or of arriving at stopping places—that occur in scientific arguments seem insurmountable at the times when they occur, these temporary points of stoppage can sometimes be exactly what is needed for an argument to advance to the next stage and result in the production of new knowledge. Thus, for instance, S. Scott Graham and Carl Herndl use stasis theory to illuminate disputes that occurred among an inter-disciplinary group of specialists in pain medicine. Their analysis reveals how intense disputes about how to define pain led to moments of "interpretive-definitive stasis," and disputes about which discipline's evidence was most suit-able for the treatment of pain led to moments of "evidential-translative stasis." In Graham and Herndl's analysis, however, these moments of stasis were ulti-mately productive because they led this group to move beyond their disciplinary differences and to "explore a hybrid discourse on pain."[2]

Stasis, in analyses such as these, is revealed as an aspect of rhetoric that is fraught with contradictions. It is a stopping point that contains movement. It is a temporary pause that does not always indicate full paralysis of the argument, but rather a place of waiting for the force to build to such an extent that it eventually has to be released, which pushes the argument in a new direction. As a rhetorical concept, then, stasis offers an intriguing next move from the previ-ous chapter's focus on topos. If topos draws our attention to the moments of breakthrough or discovery in science—that is, the moments when science seems to move suddenly forward—stasis draws our attention to the seemingly oppo-site phenomenon of moments when science appears to stand still. In the history of medical beliefs about hysteria, I argue in this chapter, the nineteenth century can be understood as just such a moment. On one hand, a desperate adherence to ancient beliefs characterized this era. On the other hand, experts in many of the newly emerging subdisciplines of medicine were striving, although not suc-ceeding, to produce explanations and cures for hysteria that would meet the era's increasing demands for scientific verifiability. In fact, hysteria received so much attention from experts in the nineteenth century that this era has been dubbed "the age of hysteria."[3] As I argue in this chapter, it is these two contradic-tory "rhetorical motions"[4]—the desire to cling to ancient ideas about hysteria, and the intense efforts to achieve a scientific explanation that would meet the needs of nineteenth-century experts—that establish the contours of stasis in nineteenth-century arguments about hysteria. In other words, these two coex-isting "rhetorical motions" tell us, in general terms, what the point is in the argu-ment "that must be settled somehow before any further progress can be made in discussing the overall problem."[5]

In addition to highlighting the different ways in which ideas move (or fail to move) through time, this chapter also draws our attention to the contrast between noise and silence as another way in which movement and stillness can manifest in scientific discourse. The previous chapter revealed the noise that was produced by the multiple, conflicting beliefs that accrued over the centuries since ancient times and amalgamated into popular texts such as the two that were analyzed in that chapter. The present chapter adds to that noise by exposing this other dimension of nineteenth-century scientific arguments about hysteria—that is, a dimension of stillness and quiet that can occur at moments when arguments reach paralysis. The chapter's specific focus is a series of lectures, published in 1877 and delivered in the years just prior to that, by French physician Jean-Martin Charcot. Now known as the "father of neurology," Charcot was part of a new breed of experts in the many new medical subdisciplines, including neurology, that were emerging in this era. He earned this nickname because, in the decades before he turned his attention to hysteria, he had identified physical causes and explanations for a number of neurological conditions that are still known today, including Tourette's syndrome (named after one of Charcot's assistants), multiple sclerosis, and amyotrophic lateral sclerosis (now known in the United States as Lou Gehrig's disease, but known in France as Charcot's disease).[6] In contrast to the popular texts that were examined in the previous chapter, the rhetoric of specialists such as Charcot was aimed at audiences of their fellow experts, whose expectations for scientific evidence underwent a steady increase between the seventeenth and twentieth centuries. In response to this increasing demand for scientific evidence, these experts sought innovative empirical approaches to understanding and treating conditions such as hysteria.

What makes hysteria an interesting case for scientific rhetoric, I argue in this chapter, is that these experts were never able to resolve the stasis that had come to characterize nineteenth-century arguments about hysteria. Instead, hysteria withered away and was replaced by other diagnoses. Nonetheless, examining the nineteenth century as a period of stasis reveals how the rhetorical activity of those who were trying to get these arguments moving again led to a new set of ideas about women's brains and bodies that outlasted hysteria itself. Specifically, as we will see in subsequent chapters, through intensifying their focus on women's brains and emotions—rather than on the uterus—as the primary location of hysteria in the female body, nineteenth-century scientists became increasingly interested in the cyclical nature of hysteria symptoms, observing that these symptoms seemed to correlate with the menstrual cycle, although they could not explain how or why. Additionally, the hysterical woman came increasingly to be seen as a menace to society, starting with her immediate family and extending to the larger social system in which she lived. In articulating these new certainties about women's bodies and, at the same time, acknowledging the new uncertainties that they created, these nineteenth-century scientists laid the foundation for some assumptions about the female reproductive organs' influence on the brain that, in subsequent decades, came to be incorporated in the concept of the hormonal woman. And, in fact, some of these concepts persist in contemporary science. In this sense, stasis can be understood both as a pause in an argument and as a moment full of great potential for the kind of movement that is crucial for an argument to advance. Examining a key moment of stasis, thus, exposes another important dimension of rhetoric as movement, offering not only a contrast to the previous chapter's focus on topos, but also, in some ways, causing us to question the rhetorical tradition's tendency to treat these moments of breakthrough and of stillness as different kinds of moments in scientific argumentation.

Change and Contradictions in Nineteenth-Century Science

Charcot's attempt to develop a scientific approach to understanding hysteria has to be understood in the context of larger changes that unfolded between the seventeenth and twentieth centuries. The seventeenth century is, of course, when the scientific article emerged as a genre, which was significant in itself. Reporting scientific findings in the peer-reviewed article format that we expect in science today implies a currency, contestability, and verifiability that is not available in popular books like the ones that were analyzed in the previous chapter.[7] There were, of course, external pressures that led to these changes. The individual experiments became more complex, the number of scientific articles multiplied exponentially, and the burden of proof dramatically increased. There was also increasing scrutiny from the readers of these texts, as "the community of science had become far less tolerant of argumentative diversity."[8] And because there was more competition thanks to the greater number of scientific articles available, scientific authors had to be more efficient in their presentation and use textual features such as titles and subheadings to draw readers' attention to important parts of their texts. These changes occurred gradually from the seventeenth through twentieth centuries. During the nineteenth century, experts began arguing with each other rather than the ancient authorities in an attempt to establish the significance of their own subdisciplines. This made the overall terrain much more complex—that is, rather than just a two-way struggle between ancient authorities and experts, we see something more like modern scientific medicine, in which the experts in different subdisciplines began competing to offer the most effective and accurate explanation of patients' health conditions.

Although the origins of modern scientific mechanisms such as peer review are tied to the seventeenth-century founding of the Royal Society of London, many of the institutions that we now recognize as integral to science experienced dramatic growth in the nineteenth century. Between 1800 and 1900, for instance, the number of scientific journals in the world rose from 100 to 10,000, and this increase in journals was accompanied by a dramatic increase in the number of professional organizations whose membership was restricted to scientists with formal training and expertise in a particular field.[9] Related to this growth in the number of venues developed explicitly for experts to communicate with each other, the nineteenth century is widely known as a time during which science became professionalized, leading to sharper distinctions between those who pursued science as a hobby and those who pursued it as a career. Along these lines, the genre of the scientific article also developed to feature less personal language and more stylistic uniformity, possibly indicating that "the science professional [was] replacing the science enthusiast in ever increasing numbers as both reader and author."[10] This was also a time when many new medical specialties originated, so the human body was starting to be seen as a collection of separate parts or systems rather than a whole.

Despite this evidence to suggest that the nineteenth century was an era of scientific professionalization, however, this fundamental change was not yet complete. Scientific publications in the nineteenth century still reveal much more blurring of boundaries between expert and nonexpert audiences than we typically see in today's scientific journals. Even the most specialized journals in the nineteenth century, for instance, typically included news articles and other nonresearch items that were aimed at a wide mix of readers, alongside the original research reports that were clearly written for fellow scientists who shared the author's expertise. In terms of style and argument, nineteenth-century scientific articles were becoming more formal than their predecessors were, but they were still far less formal and consistent than twentieth-century articles were, and authors did not yet adhere consistently to the citation practices that readers came to expect in the twentieth century.[11] As we will see, this flux and change that occurred in nineteenth-century science and medicine set the stage for the points of stasis or stoppage that Charcot and his colleagues encountered as they took up the task of developing a scientific approach to hysteria in the late nineteenth century.

Charcot as a Rhetorician

Although numerous experts in Europe and the United States devoted their attention to hysteria in the nineteenth century, this analysis begins with the contributions of Charcot. Between 1862 and 1870, in his position at Pitié Salpêtrière Hospital, Charcot established what eventually came to be known as the discipline of neurology. In the 1860s, he began conducting research that was known at the time as neuropathology: the careful study of the brains of dissected corpses to find visual evidence of the mental-health conditions that he had observed in live patients at the hospital. This careful study of brains retrieved through autopsy was an increasingly important method of study in the nineteenth century, and it was employed by researchers in many disciplines who increasingly saw the brain as fundamental to a patient's identity and wellness.[12]

In the 1870s, Charcot's focus shifted to the study of hysteria. His interest in hysteria continued to grow, and in the 1880s he began to perform public hypnosis demonstrations that presumably allowed physicians in the audience to observe hysteria patients experiencing different types and stages of hysteria attacks. Although the initial audience for these demonstrations was limited to his fellow physicians, the weekly events later developed into "a spectacle and a circus" that attracted a much wider audience, including laypeople.[13] As demonstrated in chapter 1, ideas about hysteria had been accumulating since ancient times, and although experts were suggesting many new ideas by the time when Charcot was doing this work, no one could definitively refute the ancient ideas. Thus, in the widely circulating popular texts that were the focus of the previous chapter, an assortment of ideas from ancient times to the present had come to be presented alongside each other. Readers of those texts were presented with many conflicting ideas but no clear mechanism for sifting through those ideas. In terms of stylistic features, the texts examined in chapter 1 frequently refer to older texts, and they sometimes even quote from these texts, but there is no expectation that a reader could locate those texts to verify; rather, it is expected that the author's knowledge of those previous texts will be trusted. Furthermore, in terms of subject matter, it is noteworthy that the popular texts analyzed in the previous chapter addressed many aspects of the human body without reflecting the sense of disciplinary specialization that was emerging for many scientists in the nineteenth century.

By contrast, nineteenth-century experts such as Charcot, who wanted to be on the side of the new science, devoted their energy to compiling facts and empirical evidence that they had observed with their own eyes and touched with their own hands. These experts tended to focus narrowly on one part or aspect of the body, and they wanted to present their evidence in an almost encyclopedic fashion, as if the facts would speak for themselves. In this regard, some of the texts examined in this chapter are reminiscent of the earliest scientific articles that were published when this new genre was first emerging in the seventeenth century. As Alan G. Gross, Joseph E. Harmon, and Michael Reidy suggest, the earliest scientific articles embody a view of science "as a museum that, in the ideal case, incorporates every fact about the natural world."[14] Along these lines, the expert authors of the texts considered in the present chapter avoided the speculation and theorizing that were affiliated with those who still believed in the ancient ideas of Hippocrates, Aristotle, and the like. Whereas we see frequent references to authoritative ancient texts in the popular texts that were the focus of the previous chapter, the expert texts considered in the present chapter indicate a desire to break free from the authority of ancient texts and to establish an approach to hysteria that would meet the growing demands for a science-based medicine.

To understand more precisely the points at which these experts' rhetorical efforts were stalled, it is necessary to consider the specific types of stasis that have been established in rhetorical tradition. This category scheme began with the four basic types of stasis suggested by Cicero: conjecture, definition, quality, and procedure. This basic scheme has evolved to meet the needs of rhetorical scholars in the context of specific kinds of rhetorical situations. Most relevant to this analysis, Lawrence J. Prelli has developed a two-level scheme to capture the different kinds of stasis that are typical in scientific debate. In the context of science, Prelli identifies four superior stases from Cicero's four-part scheme. Cicero's types can be translated into the evidential, interpretive, evaluative, and methodological elements of scientific work. Prelli then identifies four subordinate stases that are meant to divide the superior stases further into subtypes. He presents these types of stases in a grid that ultimately presents sixteen categories or subtypes,[15] as shown in table 1.

Table 1   Rhetorical Stasis Procedures of Scientific Discourse

Superior Stases

Subordinate Stases	Evidential	Interpretive	Evaluative	Methodological
Conjectural	Is there scientific evidence for claim x?	Is there a scientifically meaningful construct for interpreting evidence?	Is claim x scientifically significant?	Is procedure x a viable scientific procedure in this case?
Definitional	What does the evidence mean?	What does construct y mean?	What does value z mean?	What does it mean to apply procedure x correctly?
Qualitative	Which empirical judgments are warranted by available evidence?	Which interpretive applications of construct y are more meaningful?	Which claims are more significant, given value z?	Which investigations exemplify appropriate applications of procedure x?
Translative	Which evidence more reliably grounds claims about what does and does not exist?	Which scientific constructs are more meaningful?	Which scientific values are more significant?	Which procedures more usefully guide scientific actions?

Source: Reprinted from Lawrence Prelli, *A Rhetoric of Science*, with permission from University of South Carolina Press.

In Prelli's scheme, the points of stasis that appear in nineteenth-century arguments about hysteria fall into the categories of evidential stasis and interpretive stasis; that is, they revolve around questions about the kinds of data that are available, the appropriateness and legitimacy of that data, and how the data should be interpreted. The three specific types of stasis that characterized scientific arguments about hysteria in the nineteenth century form the basis of the structure of the analysis below.

Interpretive-Definitional Stasis

In the terms established by Prelli's scheme, an interpretive-definitional stasis arises from questions that take the form of the question, "what does construct y (hysteria) mean?"[16] More specifically, these scientists were asking whether hysteria is an actual physical disease that can be traced to a defect in the brain or

uterus, or whether we have to explain it in another way. One of the two components of this type of stasis is interpretive stasis, which "becomes prominent
when scientists accept sets of data as facts but have difficulty deciding what
theoretical constructs or models accommodate them."[17] The second—defini-
tional stasis—is a subtype of interpretive stasis that emerges "when the meanings of constructs are at issue" or when "principles used to develop classification
schemes are questioned or are otherwise confused or ambiguous."[18] When
Charcot was establishing himself as a hysteria expert in the 1870s, these were
the very kinds of stoppage points that experts were encountering as they confronted many centuries' worth of differing beliefs that had accumulated about
the etiology of hysteria. These experts, working from various subdisciplines of
medicine, largely agreed that hysteria existed. However, how they defined the
disease and its etiology varied greatly. Could hysteria have been traced to a
defect in the body—whether in uterus, brain, or something else? Or was it more
like what we understand today as a mental illness—that is, an emotional disturbance that could not be traced to a physical defect in the body? These are some
of the questions that motivated Charcot's work and formed the stasis of the
scientific arguments about hysteria into which he was entering.

Historians of medicine generally concur that by the late eighteenth century,
the neurological understanding of hysteria had replaced earlier uterine theories
of the disease; in other words, most experts who were seeking a scientific understanding were more inclined to perceive hysteria as a disease of the brain than of
the uterus.[19] To reiterate briefly from the previous chapter, this shift toward
emphasizing the brain, rather than the uterus, as the disease site in hysterical
patients began in the seventeenth century with physicians who tried to refute
witchcraft and demonic possession as the causes of the disease. Most notably,
Jorden was one of the first experts to speculate—in his 1603 treatise, *Briefe Discourse of a Disease Called the Suffocation of the Mother*—about the brain's role in
hysteria.[20] Although he is widely known as the first physician to acknowledge
the brain's role in hysteria, Jorden did not deny the womb's involvement in the
disease.[21] The first clear assertion that hysteria originated in the brain came
later, from Charles Lepois in a 1618 publication.[22] Whereas Jorden acknowledged
the brain's involvement, Lepois was the first to refute the uterine theory of hysteria and declare that hysteria originated in the brain, with effects flowing from the
brain to the other parts of the body rather than moving in the opposite direction. The next important author, Willis, denied the uterine theory of hysteria;

he allowed that the uterus could have sometimes been involved, but as an early neurologist, he focused much more on the brain's involvement—specifically, Willis believed that "animal spirits" of the brain were the cause of hysteria.[23]

The earliest physician who is seen as having a direct influence on Charcot's approach to hysteria was seventeenth-century physician Thomas Sydenham, who agreed with experts who emphasized the brain's role in hysteria. He distinguished himself from many other experts by claiming to offer a clinically based explanation of hysteria symptoms, which included a description of a number of mental symptoms that sound close to what we today might call depression. He did not seem to locate the cause of hysteria either in the uterus or the brain; rather, he saw it as existing in the body as a whole. Sydenham also echoed Willis in asserting that hysteria resulted from "disturbances of the (animal) spirits." Thus, hysteria was not strictly a brain or uterine disorder for Sydenham; instead, "it is a disease of the integral person."[24]

Because the neurological understanding was more conducive to scientific study than had been the case for previous, womb-based explanations, this shift to a neurological understanding was seen by many experts at the time as an indication of scientific progress. By the late eighteenth century, however, a frustration had developed with the lack of progress in finding new treatments for hysterical patients. Physicians throughout the seventeenth and eighteenth centuries continued to recommend old cures, but they adapted the explanations for these cures to fit the new theories of the disease.[25] Thus, for instance, we can see evidence in Sydenham's language of a divide between his therapies for hysteria and his descriptions of the disease's etiology. He relied on many of the older therapies, such as blood purification and horseback riding, but he matched them to his new explanation of the disease's etiology. For instance, he recommended the old practice of bleeding and purging to purify the blood and rid it of the bad humors, and he recommended horseback riding, which had long been recommended as a cure for hysteria. But Sydenham's explanation of these cures was tied to his understanding of hysteria as a mental disease, so he justified his recommendation of horseback riding by explaining that exercise was good for the mind, not because it would jar the womb back into its proper place as earlier experts had suggested.[26]

Although physicians throughout the eighteenth century continued to seek new innovations, they also remained largely dissatisfied with the lack of progress in treating hysteria. Ilza Veith asserts that this "duality of science and tradition-bound empiricism in therapy" emerged in seventeenth-century treatments for hysteria.[27] Clearly, as my analysis suggests, this tendency persisted well into the eighteenth century. This lack of progress in finding new treatments for hysteria existed simultaneously with so many centuries of experts' proclaiming to offer novel understandings of the disease, and supported a characterization of the period from the late eighteenth century to the early nineteenth century as a key moment of interpretive-definitional stasis in experts' arguments about hysteria. One eighteenth-century attempt to break free from the constraints that became increasingly evident in neurological explanations of hysteria is seen in the work of Phillippe Pinel, one of the physicians who is responsible for establishing psychiatry as a discipline in the late eighteenth and early nineteenth centuries.[28] In 1794, Pinel was appointed to be in charge of the Salpêtrière. Early in his tenure there, he published two major treatises on his ideas of liberating mentally ill patients by seeking new treatments for their conditions, rather than continuing to treat them as prisoners. One of Pinel's treatises was on mania, which eventually evolved into a complete textbook on psychiatry. Pinel's assertion that mental diseases could be cured was revolutionary: He refuted the long-standing notion that mental diseases were caused by organic lesions in the brain. If a disease were seen as the product of an organic defect, it would have been seen as incurable, but if it were seen as more of a mental or emotional disturbance, then it would have been subject to therapy. Reflecting his belief in the latter of these two explanations for hysteria, Pinel started trying some early versions of psychotherapy to cure his patients. He developed a classification of diseases, placing hysteria under "genital neuroses of women," a category that also included nymphomania, frigidity, and sterility. Pinel is known for the novelty of his understanding of hysteria as a mental condition—that is, as a condition that did not originate in a physical defect in either the brain or the uterus, but rather was a condition that arose from a mental or emotional disturbance. Despite this apparently novel approach, however, the late eighteenth and early nineteenth centuries were a moment of standstill in scientific advancements for understanding hysteria. As Veith says, by this time, "the vague neurological hypotheses for hysteria had been repeated so often without the addition of any new thought that they had become practically meaningless."[29] We might say that by the late eighteenth century, womb-based explanations had given way to brain-based explanations, but because both had proven to be dead ends on the road toward effective treatments for hysteria, expert arguments at this time had reached a standstill.

Charcot's rhetoric indicates his desire to dissociate himself from the old way of framing the debate between those who saw the womb as the site of hysteria and those who more recently had come to see the brain as the site of hysteria. Charcot was influenced by earlier physicians such as Sydenham and Pinel, but he sought to move beyond the standstill in these physicians' efforts to find new explanations of hysteria that would lead to innovative cures. Thus, Charcot explicitly rejected sexual explanations of hysteria—such as those that traced hysteria to the womb—but he also distinguished himself from Sydenham's and Pinel's explanations by articulating the novel idea of the "ovaries as an hystero-genic zone."[30] In positing this explanation for hysteria, Charcot went outside neurology and drew from another newly emerging field that would, a few decades later, come to be known as endocrinology. Specifically, Charcot built on the work of Charles-Edouard Brown-Séquard, who later came to be known as the father of endocrinology. Charcot adapted Brown-Séquard's idea of "epilep-togenic zones" and tried to identify similar "hysterogenic zones, the points in the female body which when pressed could set off an hysteric crisis."[31] Extending the analogy to Brown-Séquard's epileptogenic zones, Charcot claimed that dur-ing the "aura hysterica" phase of a hysteria attack, compression of the ovaries had a therapeutic effect; Brown-Séquard had made a similar claim about compres-sion of the epileptogenic zones during the epileptic "aura."[32]

One of Charcot's published lectures, "Ovarian Hyperaesthesia," elaborates upon his theories about the ovaries' involvement in hysteria and documents several cases in which ovarian compression proved successful as a cure. In this lecture, Charcot's positioning of his own explanation of hysteria in relation to past experts' uterine explanations of the disease offers important insights into interpretive-definitional stasis in the argument into which Charcot was enter-ing. First, he says about ovarian compression that it can be traced to the ancients: "The invention of this process is far from being my own; it may possibly be traced to a very ancient period; it is certain that it dates from a time anterior to the sixteenth century."[33] Next, he discusses several past experts—starting with Willis in the seventeenth century—who had recommended a similar technique of ovarian compression to treat hysteria. He cites texts to support his claims about these previous experts, using footnotes and complete citation informa-tion, in a manner that resembles citation practices of modern science, rather than the haphazard citation practices that are more typical in premodern scien-tific texts.[34] Charcot also explicitly distinguishes his beliefs from those of previ-ous experts who traced hysteria to the womb instead of the ovaries, characterizing these ideas as outmoded and unscientific: "But Mercado (in 1513) had long previously advised the use of *frictions on the abdomen*, with the object of reducing the womb, which he supposed to be displaced, according to the old doctrine."[35] Thus, even as he tries to offer innovative ideas that will unsettle the stasis in this era, he acknowledges the ideas of previous experts on which his own ideas clearly build; he maintains a delicate balance between aligning his own ideas with elements of the older ideas and dissociating his ideas from elements that seem problematic.

Charcot's work offers a glimpse of a new understanding of hysteria that does not posit any direct link to physical problems in the brain or the uterus. For Charcot, and a few of his predecessors, we see early versions of an idea that would become more pronounced later: that hysteria is more of a mental or emotional disorder than a physical disease. However, Charcot insisted throughout his career that hysteria "was a genuinely organic disorder, a disease rooted firmly in the higher nervous system, and in these respects part of the broader spectrum of neurological disorders."[36] Although he could not identify the kinds of lesions and physical defects that he had discovered in corpses that had suffered from other neurological disorders such as Parkinson's disease and multiple sclerosis, he continued to espouse this belief that hysteria had some physical component. Interestingly, even though he insisted that hysteria was a neurological disorder, the closest he came to offering a specific disease location was in his emphasis on the ovaries as hysterogenic zones. With this idea, he started to unsettle the stasis that had formed after several centuries of experts' tracing of hysteria to either the brain or the womb. In Prelli's scheme, the particular stasis surrounding this aspect of the debate is most accurately characterized as interpretive-definitional stasis because it involves questions about how to explain hysteria as a disease and where to pinpoint its origins in the body.

Evidential-Conjectural Stasis

Interpretive-definitional stasis, which arose from questions about the etiology of hysteria, also led to other forms of stasis, including an evidential-conjectural stasis, which arose from the question of whether hysteria could be scientifically diagnosed. In this section, we see that although experts largely agreed that hysteria exists, Charcot faced the rhetorical challenge of proving that it could be diagnosed in a way that would meet increasing expectations for scientific evidence. In Prelli's scheme, evidential stases occur "whenever issues arise concerning the scientific relevance and situational appropriateness of empirical evidence" and of judgments related to that evidence that are "used to address questions about existence." Furthermore, in Prelli's scheme, "conjectural stasis occurs in evidential scientific discourse whenever there is ambiguity about the availability or reliability of evidence."[37] An evidential-conjectural stasis, then, arises from questions such as whether there is scientific evidence for a given claim. More specifically, in this context, the stasis arises from the question of whether hysteria can be scientifically diagnosed with the same precision that other diseases can be, or whether it is an outmoded diagnosis that is confined to the realm of superstition and ancient myths.

In other words, experts studying hysteria at this time were asking, can hysteria live up to the new standards of science in the nineteenth century, or will it always elude these efforts, remaining the mysterious disease that it had been for so many centuries? Does hysteria truly exist as a disease that is worthy of scientific study, or is it just a condition sustained by myths and misperceptions? These questions were not meant to deny the existence of hysteria as a condition that people (mostly women) had been experiencing for centuries, but to argue about whether, as a disease, it could be subject to scientific understanding. After many centuries of experts' theorizing about hysteria, with speculations that derived from the ideas and observations that were documented in the ancient texts, the increasing number of experts who identified with the new subdisciplines of medicine that were taking shape in the nineteenth century put pressure on these old explanations, which were no longer adequate in a scientific world that demanded clinical evidence and precision.

Many of Charcot's French colleagues were quite skeptical that theories about hysteria could ever live up to the new standards of a scientific medicine.[38] Charcot responded to these questions about the scientific legitimacy of hysteria with relentless effort—one of the defining features of his scientific work—to document the stages and symptoms of hysterical attacks and to differentiate the various forms of hysteria with which patients could be diagnosed. According to historian Diana P. Faber, an important part of Charcot's agenda was to redefine several previously known disorders as manifestations of hysteria. Even "behaviours given religious or other explanations in the past could now be subject to what Charcot believed to be scientific scrutiny."[39] One might say that Charcot's intent was to medicalize hysteria. That is, he tried to refute those who said it was just a disease of modern times, limited to certain geographic locations, or the product of demonic possession. Charcot sought instead to characterize hysteria as a biological affliction, possibly akin to how depression now is seen by the medical community as a biological condition that can be treated as such. One of the ways in which he did this was through "retrospective diagnoses" of patients who were previously believed to have experienced "extreme religious experiences, ecstasy and hallucinations."[40]

As evidence to support his conviction that hysteria could be diagnosed, he provided numerous photographs of hysterical women during specific phases of the attack (figures 5 and 6). He also engaged in public hypnosis demonstrations that presumably allowed audience members to observe the hypnotized patients experiencing hysterical attacks. The use of visual evidence was important for Charcot, and he is known in the history of medicine largely because of his extensive photograph collection and his public hypnosis demonstrations. Many historians have emphasized the rhetorical impact of Charcot's photographs, which, of course, "carried the illusion of providing the truth . . . the instantaneous representation of what passed before the lens of the camera," but were also the product of elaborate staging and manipulation that were obscured in the final versions that have been preserved for historical record.[41] Faber suggests that the visual evidence that Charcot provided was crucial because nineteenth-century physicians struggled with the verbal elements of the disease. As she says, "it is possible that [Charcot's] preference for the visual elements of hysteria, in demonstrations, drawings and photographs led Charcot to overlook, to some extent, the problems involved in verbal description."[42]

Charcot's work with hysteria patients, including the public hypnosis demonstrations and accompanying lectures, enabled him to open for public scrutiny the hysteria behaviors and symptoms that fascinated him. Although some male patients were among the hysteria cases that he treated, most of his patients were women, and the distinctly gendered dimension to the concept of hysteria is part of Charcot's legacy. By identifying the various stages of the hysterical attack, and providing photographic evidence to show patients experiencing these stages, he aimed to convince his fellow experts that hysteria could be more than the mysterious, womb-based ailment that it was depicted as being for centuries since the ancient texts. Thus, we see in his lectures a recurring emphasis on the visible and even tangible nature of the evidence that he presented. In one of his published lectures, he offered one vivid example to illustrate how he used these rhetorical techniques: "This pain I shall enable you to touch, as it were, with the finger, in a few moments, and to observe its characteristics, by introducing to your notice five patients who constitute almost the whole of the hysteria cases,

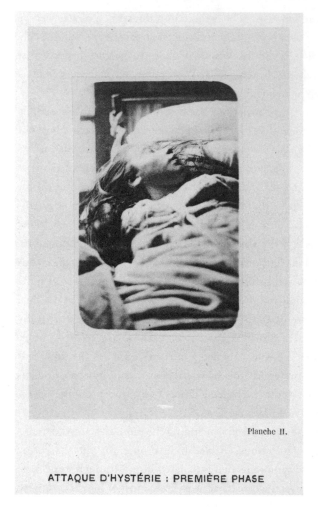

Planche II.

ATTAQUE D'HYSTÉRIE : PREMIÈRE PHASE

5 | "Attaque d'Hystérie—
Première Phase." From
*Iconographie photographique
de la Salpêtrière* (Service de
M. Charcot), by Désiré-
Magloire Bourneville and
Paul Régnard, vol. 1, 1876–
1877 (Paris: Aux bureaux du
Progrès medical, V. Adrien
Delahaye and Cie., 1876).

actually existing among the 160 patients who occupy the department devoted in
this hospital to women affected by incurable convulsive diseases, and reputably
exempt from mental alienation."[43] Reflecting the idea of the nineteenth century
as a moment of stasis, or standstill, in these ongoing arguments, we can see in
passages such as this an intense effort to achieve a precise definition and understanding of this disease by medical experts. Charcot was "well known for his
ability to make fine distinctions in his clinical observations."[44] In this regard, his
approach to hysteria was influenced by the seventeenth-century work of Sydenham and the nineteenth-century work of Paul Briquet. He followed Briquet in

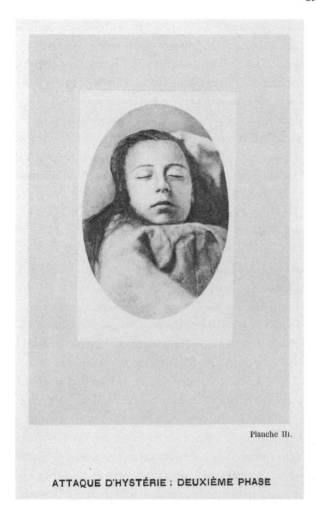

Planche III.

ATTAQUE D'HYSTÉRIE : DEUXIÈME PHASE

6 | "Attaque d'Hystérie—
Deuxième Phase." From
*Iconographie photographique
de la Salpêtrière* (Service de
M. Charcot), by Désiré-
Magloire Bourneville and
Paul Régnard, vol. 1, 1876–
1877 (Paris: Aux bureaux du
Progrès medical, V. Adrien
Delahaye and Cie., 1876).

the attempt to use the scientific method to find a more precise way to diagnose hysteria. Briquet had said in an 1859 publication that hysteria "obeyed laws that could be determined; that the diagnosis could be made with as much precision as other diseases." Traces of this idea from Briquet are evident in Charcot's "reference to the possibility of making a precise diagnosis of hysteria by the identification of its laws, for Charcot's search for laws was in accord with his belief in a new scientific method."[45]

However, Charcot also explicitly dissociated himself from Briquet in many places throughout his lectures—further proving that Charcot was attempting to unsettle the stasis of nineteenth-century arguments about hysteria—and he used the rhetorical technique of bringing tangible evidence to his audience as a way to do this. He says in one lecture, for instance, "it now becomes our duty to examine how far we ought to follow this eminent author [Briquet] in the path which he has marked out for us."[46] Almost as if he were walking the audience along with him on this path, Charcot then describes in great detail how he worked with actual ovaries and surrounding organs from dissected corpses to pinpoint the specific location of this ovarian pain that he says is the seat of hysteria. After this detailed verbal description, which also includes a reference to a diagram that shows "a horizontal transverse section of the body of a woman, aged 20,"[47] Charcot reiterates his conviction that the ovary is the site of hysteria's origins and notes this as a point of disagreement with Briquet:

> Gentlemen, after all these explanations which I have just discussed, I believe I have a right to draw the conclusion that it is to the *ovary*, and the *ovary alone*, we must attribute the *fixed iliac pain of hysterical patients*. It is true that, at certain epochs, in severe cases, the pain, by a mechanism which I need not at present indicate, extends to the muscles and to the skin itself, so as to justify the description given by M. Briquet; but I cannot too often repeat that, if limited to these external phenomena, the description would be incomplete and the true focus of the pain misapprehended.[48]

When Charcot characterizes Briquet's judgments as based on "external phenomena," he is saying that his own evidence is more meaningful than Briquet's because it is based on anatomical evidence retrieved from a dissected corpse. He acknowledges that Briquet might have interpreted this ovarian pain differently, but he contends that his own interpretation is more trustworthy because it is based on anatomical evidence from a dissected corpse, whereas Briquet's interpretation is based strictly on manual examination of live patients.

In addition to the anatomical evidence that he had retrieved from dissected corpses, Charcot also, in some places throughout these lectures, refers to live patients who are in front of the audience. Based on what we know about Charcot's rhetorical practices, these were patients who had most likely been hypnotized in a way that caused them to appear as if they were experiencing hysteria attacks in front of the audience. This was valuable to Charcot because it allowed him to demonstrate techniques such as ovarian compression in front of his audience. For instance, in the following passage, he narrates a demonstration of the technique of ovarian compression as a remedy for hysterical attacks:

> You have just seen how methodical compression of the ovary can determine the production of the aura, or sometimes even a perfect hysterical seizure. I propose now to show you that a more energetic compression is capable of stopping the development of the attack when beginning, or even of cutting it short when the evolution of the convulsive accidents is more or less advanced. This, at least, is what you can very plainly discern in two of the patients whom I have placed before you. In their cases, the arrest of the convulsion, when compression has been properly applied, is total and final. In the others, the manipulation merely modifies the phenomena of the seizure in varying degrees, without however, producing complete cessation.[49]

In the last few pages of this section, Charcot then presents detailed case studies of the individual patients to whom he has been referring throughout this lecture, and he uses these "to point out the most salient peculiarities which they present." In discussing each patient, he talks about their symptoms and how long they have experienced them, and he documents the effects of ovarian compression. In some cases, he also provides details about their lives and mentions these previous experiences as possible causes of hysteria. For example, in regard to one patient, he says, "the ill-treatment she had suffered from her father, who was addicted to alcoholic excesses, and her subsequent career as a prostitute, have doubtless exerted a certain etiological influence."[50] The printed versions of these lectures include line drawings to illustrate what some patients look like during various stages of the attack, replicating patients depicted in Charcot's photographs.

This glimpse into Charcot's rhetorical tactics suggests some of the reasons why hysteria was dubbed the "mocking bird of nosology" in 1849.[51] It was clearly an important part of Charcot's life mission to make hysteria fit into some kind of predictable scheme, and he faced several challenges in accomplishing this goal. The difficulties that hysteria presented for nineteenth-century physicians stemmed from "its protean and polysymptomatic nature and its perceived ability to simulate other disorders."[52] In an era that increasingly demanded scientific precision and strict classification schemes, hysteria was slippery, hard to define; this difficulty was compounded by the fact that hysteria was frequently confused with other diseases such as epilepsy, a point that will be elaborated in the next section. All of these traits suggest some of the reasons why hysteria was such an object of intense focus, attention, and frustration for experts such as Charcot who were trying to find a way to make this disease live up to the standards of nineteenth-century science. The evidential-conjectural stasis emerging from the questions about evidence explored in this section was clearly an important part of the rhetorical exigency that movitated experts such as Charcot.

Evidential-Definitional Stasis

After the question of whether there is scientific evidence of hysteria, the next stasis or stopping point that is visible in arguments about hysteria emerges from the question, what does the evidence mean? More specifically, this stasis arises from questions such as whether a physician's observation of the symptoms of hysteria are considered legitimate scientific evidence for the diagnosis of hysteria, or whether another legitimate explanation for what was called hysterical behavior exists. Whereas an evidential-conjectural stasis arises from questions about the availability or reliability of evidence, an evidential-definitional stasis arises from questions about how to assign meaning to that evidence. In Prelli's words, an evidential-definitional stasis arises "when there is ambiguity about what the available evidence means."[53] Along these lines, an important focus of physicians' debates was the question of how to interpret the evidence that Charcot presented. In other words, is the legitimacy of the illness truly demonstrated by Charcot when these hysterical women acted out their hysterical attacks while hypnotized in front of a public audience, and by their being included in Charcot's collection of photographs? Or could it mean that they were acting up and imitating other disorders such as epilepsy to gain attention? If it were the former, then hysteria could be seen a legitimate disease with its own defining characteristics; if it were the latter, then these patients' behaviors had to be understood in a different way. They may have been experiencing a specific kind of epileptic seizure, for instance, that was misdiagnosed as hysteria, or they may have been imitating the behaviors of patients who were truly ill, and they may have been doing so simply to gain attention.

These possibilities for multiple interpretations become especially apparent in some of Charcot's photographs. For example, the collection includes a number of photographs in which it is clear that the subject has not been a passive object, but rather has posed in a specific way for the purpose of the photograph. As shown in figure 7, the subjects in these photographs often assume poses that have erotic overtones. By contrast, other photographs in the collection depict women staring blankly without appearing to be in an intentional pose of any sort. The photographs shown in figure 8 suggest that, in some cases, these hysteria patients were probably experiencing neurological conditions such as epilepsy that can now be accurately diagnosed and treated.

This discussion of the differences between hysteria and other diseases is a theme that can be traced to ancient times, but it took on a distinct, new shape in the nineteenth century as the clinic and clinical observations became increasingly accepted as the normal venues in which medicine was practiced. As Francis

Planche XXIX.

HYSTÉRO-ÉPILEPSIE

7 | "Hystéro-Épilepsie—Contracture." From *Iconographie photographique de la Salpêtrière* (Service de M. Charcot), by Désiré-Magloire Bourneville and Paul Régnard, vol. 1, 1876–1877 (Paris: Aux bureaux du Progrès medical, V. Adrien Delahaye and Cie., 1878).

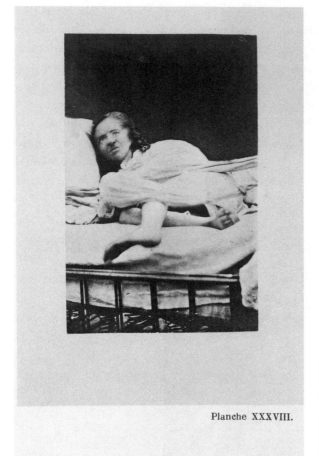

Planche XXXVIII.

HYSTÉRO-ÉPILEPSIE : CONTORSIONS

8 | "Hystéro-Épilepsie—
Contorsions." From
*Iconographie photographique
de la Salpêtrière* (Service de
M. Charcot), by Désiré-
Magloire Bourneville and
Paul Régnard, vol. 1, 1876–
1877 (Paris: Aux bureaux du
Progrès medical, V. Adrien
Delahaye and Cie., 1876).

Schiller observes, "hysteria had been a clinical entity since antiquity precisely because those somatic manifestations were never considered to be quite as 'real' as those found with, say, strokes or other brain diseases."[54] This question of hysteria as real somatic symptoms versus hysteria as imagined mental symptoms is addressed in earlier texts such as Sydenham's, but it became crystallized in the mid-nineteenth century as a debate between those who continued to imagine hysteria as a uterine (somatic) disease, relying largely on the authority of ancient texts, and those who sought increasingly to find evidence that would help them understand it as a disease that involves the central nervous system, and could therefore be treated through newly emerging techiques in neuroscience and neuropsychiatry.

For instance, Charcot was especially interested in distinguishing between epilepsy and hysteria and in reversing the historical tendency to conflate these two conditions, which had led patients with both conditions to be housed together at the Salpêtrière when Charcot began his position there. Charcot thought it very important to achieve a distinct definition of hysteria, separate from epilepsy. Nonetheless, his initial position (from 1867 through the mid-1870s) involved work with epileptic patients who did not have a diagnosis of hysteria. Faber says the following about Charcot: "One of the features of his personality was surely persistence, a reluctance to accept incoherence, and an intolerance of ambiguity."[55] Willis had addressed the discussion about the relationship between epilepsy and hysteria in his seventeenth-century texts, in which he says that both diseases emerged from the animal spirits.[56] But Charcot differed from these earlier experts because he relied on a method that French medicine became famous for, the "médicine d'hôpital," which involved the observation and accurate description of patients' symptoms. Charcot believed that "the medical problem should precede the physiological explanation—in other words, that diagnosis should be pathology-led." Furthermore, "in his teaching, Charcot favored a method of comparison; for example, a close physical examination of one patient would be followed by that of others, and as a result of his penetrating gaze Charcot would point out similarities and differences to be taken into account when making a diagnosis." His reliance on types, which we see in the photography collection and his detailed writings, was an important part of his method and his embrace of the new scientific approach to treating hysteria because it allowed him to perpetuate the idea of laws or predictable patterns that he believed could be detected in his clinical population of hysteria patients. This allowed him to "go beyond the reliance on descriptive morphology and the data of observations." He believed in universal laws that govern everything—inside the hospital and outside—and this "allowed him to incorporate epileptic phenomena into this whole or type."[57]

In 1872, Charcot "formulated his first definite diagnosis of hysteria as an entity distinct from epilepsy."[58] Charcot defined the "classic case," dividing the hysteric episode into distinct phases. William Gowers was another physician who focused on hysteria and epilepsy in the late nineteenth century. He treated epileptic and hysterical patients in his post at National Hospital for the Paralyzed and Epileptic in London. Whereas Charcot sought to establish a sharp distinction between epilepsy and hysteria, Gowers allowed for more fluidity between the two disorders. Furthermore, Charcot was biased toward hysteria when diagnosing borderline cases, whereas Gowers was biased toward epilepsy.

In his post at the Salpêtrière, Charcot was assigned to a group of patients who were diagnosed with either epilepsy or hysteria, and at the time, not much careful attention was given to differentiating between the two diseases. Both types of patients were understood in terms of their "attacks." Both conditions were subject to many myths, and they also encompassed several other poorly understood conditions, including migraines, angina, and somnambulism. Epilepsy was also seen as, at least in some patients, a form of insanity, or "transitory madness." The term *neurosis* also started to be used to characterize epilepsy, and this condition was seen as "a moral as well as a physical form of degeneracy." Hysteria and epilepsy were grouped with some other conditions that were coming to be seen as hereditary and affiliated with "degeneracy."[59]

Faber's historical analysis refers to a "confusion in terminology" between hysteria and epilepsy, and Faber says that this confusion allowed Charcot to "describe a form of hystero-epileptic attack as 'hystérie.'"[60] In an 1877 lecture titled "Hystero-Epilepsy," Charcot offers important insights into how he conceived of the distinction between hysteria and epilepsy, and the evidential-definitional stasis that surrounded this subject at the time. He begins the lecture by referring to a number of recent cases that audience members presumbly had been given the opportunity to observe if they had attended his most recent public hypnosis demonstrations. He suggests that these hysteria patients, during their attacks, sometimes displayed symptoms that resembled epilepsy. As he says, "the convulsive form of disease which is found in all these cases, is that which has been designated, in these latter times, by the name of *hystero-epilepsy*." He then proceeds to articulate what he perceives to be the commonly accepted definition of this term at the time: "It signifies that in patients, so affected, hysteria is present in combination with epilepsy, so as to constitute a mixed form, a kind of hybrid composed half of hysteria, half of epilepsy." The goal of his lecture, however, is to call into question this accepted definition and to promote his own view that hysteria is a distinct condition. Toward this end, he first acknowledges that not everyone finds the hybrid definition to be legitimate: "the camp of its adversaries still reckons many adherents." Charcot clearly aligns himself with these "adversaries." Although he acknowledges that a single patient might be afflicted with both conditions, his lecture is consistent with the long-standing agenda for which Charcot became well known: to designate hysteria as a condition that is separate from epilepsy. Further evidence for the existence of evidential-definitional stasis comes from Charcot's question to his audience: "What, then, according to this view, is the signification of those attacks, the existence of which is so clearly established by the very cases that form the foundation of our study, and in which epilepsy seems mixed up with the ordinary symptoms of convulsive hysteria?" He then states his own thesis: "The convulsion, epileptic in form, would here appear . . . as an accessory element, without altering in anything the nature of the original disease."[61] In other words, he suggests, these patients' primary affliction is hysteria, but a form of hysteria in which the attacks resemble epileptic seizures. He then cites several "competent authorities" who share his view, including, in some cases, direct quotations from their published texts. These references include footnotes that provide specific citation information, so that a reader could look up the text and verify Charcot's citations; this contrasts sharply with the texts examined in the previous chapter, and it is one feature that distinguishes Charcot's lectures resembling the modern scientific article in terms of generic conventions. In contrast to those texts, Charcot's text suggests that the author himself is not necessarily expected to be taken at face value; if readers are skeptical, as they should be, they can consult the texts that Charcot is citing, and they can consult the public evidence that Charcot has provided in his demonstrations. He closes this section with the following passage: "Be good enough to remark, gentlemen, that this is something more than a mere question of words; it is also a question of nosology, and consequently, a question of diagnosis and of prognosis. These circumstances will I trust, suffice to justify in your eyes the details on which I am obliged to enter, in order that the conviction which I entertain may take its place in your minds."[62] As this passage indicates, in addition to citing these previous experts who share his views, the scientific evidence to which Charcot refers throughout this discussion appears to be a recent series of demonstrations in which he publicly hypnotized hysteria patients, causing them to undergo attacks that could be observed by those in attendance. He calls on his audience to view this evidence with their own eyes, but in the proceeding section, he also offers extensive support to convince his audience to share his own interpretation of the evidence.

Charcot then provides numerous details, often referring to specific patients, to support this distinction between hysteria and epilepsy. One of the key distinguishing features, he says, was that hysteria could be relieved by ovarian compression, and this was never the case for epilepsy.[63] On this point, he suggests, the audience could be assured by previous experiments (citing another one of his lectures as textual evidence). Another key distinguishing feature was that the patient's intellect was never truly affected by hysterical attacks, even if these were frequent and numerous, whereas an epileptic patient would have suffered neurological damage. His next distinguishing feature, which consumes several pages, pertains to temperature changes. He provides extensive numerical evidence, documenting patients' body temperature changes in a series of charts and graphs. Through this copious evidence (which also includes a lengthy footnote-like excerpt in which he refers to specific patient cases), Charcot makes his most well-documented claim about the distinction between epilepsy and hysteria. He documents that when patients had several epilepsy attacks in a row, their body temperature rose, and the outcome was eventually fatal. By contrast, when the condition was pure hysteria, there was no temperature rise, and the patient never died as a result of these attacks.

Charcot also faced the challenge of refuting those who contended that his hypnotized hysteria patients were simply play-acting, competing with each other to put on dramatic displays of hysterical behaviors because they knew this would gain Charcot's favor and attention. A rare first-hand account written by Jane Avril, who was hospitalized at the Salpêtrière from 1882 to 1884, includes her description of the atmosphere as one of intense competitition among the large number of female patients who were hospitalized under Charcot's care and competed for his attention by acting as model hysteria patients.[64]

In sum, we might say that this evidential-definitional stasis has two parts: distinguishing hysteria from other real conditions such as epilepsy and from play-acting. Both parts pertain to the challenge of distinguishing hysteria from other possible explanations that could have been used to tell Charcot's audience members what the observed evidence meant. It seems commonly accepted among historians that he was really observing multiple forms of epilepsy, but he created a categorical scheme that was used to interpret some of these patients as hysterical. This systematic effort to observe and categorize folded into Charcot's elaborate system of categories and subcategories that were used to make sense of these various neurological conditions. In one of his lectures, for instance, Charcot outlines five phases that characterize a typical hysteria attack.[65] Laying out a scheme such as this enabled him to position the hystero-epileptic attack as one phase that occurred only in one particular type of hysteria. The most likely explanation seems to be that Charcot's post at the Salpêtrière gave him access to patients who were suffering from various forms of epilepsy and related neurological conditions that caused seizures. This coincided in an important way with the attention given to hysteria by experts throughout many centuries. By observing these hysteria patients as they experienced various types of convulsions and seizures, physicians finally had access to empirical observations of a wide array of symptoms that previously could only be imagined or were the product of folklore such as we see in *Aristotle's Masterpiece*.

Moving Beyond Stasis

Tracing the development of arguments about hysteria offers an especially interesting opportunity to observe the many changes that occurred in nineteenth-century science and the deep contradictions that characterized knowledge production in this era. There are, of course, multiple ways in which a stasis such as that which characterized these nineteenth-century arguments about hysteria can be resolved. As Prelli says, when one of the "contrary rhetorical motions" that constitute a stasis "is advanced by argument and evidence sufficient to break opposing stands, the point of stasis is resolved and further progress can be made." However, in some situations, "the point of stasis [can be] altered by common agreement to waive the issue."[66]

In the case of hysteria, the latter characterization is probably more accurate in that, by the mid-twentieth century, hysteria had vanished as a diagnostic category across the numerous subdisciplines of medicine that had emerged as new sites for studying diseases like hysteria in the late nineteenth and early twentieth centuries. Historians have offered varying explanations for this eventual disappearance of a disease after so many centuries of intense expert attention to it. One common explanation is that hysteria's nineteenth-century heyday was the result of the extreme repression that occurred during the Victorian era, and when this phenomenon subsided, the disease subsided as well.[67] Another explanation that has developed more recently is that as the psychological sciences developed throughout the twentieth century, people gained access to a wider vocabulary for expressing their fears and anxieties, and this has made the somatic manifestations that characterize hysteria less necessary and less effective as a means of responding to trauma.[68] In the decades immediately after Charcot's work gained notoriety, Sigmund Freud became known as the first to develop a psychological understanding of hysteria, which also became an important basis for his establishment of psychotherapy as a treatment for hysteria and other conditions. Having studied under Charcot at the Salpêtrière between 1885 and 1886, Freud proclaimed that he was deeply impressed by Charcot's work with hysteria patients, even though Freud's psychological explanation departed in significant ways with Charcot's insistence that hysteria was a neurological disorder.[69] After several years of intense attention to hysteria just before the turn of the century, however, even Freud abandoned hysteria in the early twentieth century. Instead, he replaced it with neurasthenia and, later, hypochondria, and referred to these as belonging to the larger category of "anxiety neuroses."[70]

Paying attention to these shifts in the labels assigned to the collection of symptoms that had long been affiliated with hysteria opens the possibility of a rhetorical interpretation of its decline as a disease. Along these lines, historian Mark S. Micale traces the decline of hysteria to the early twentieth century and contends that it was "effectively complete by World War I." Micale offers a metaphorical explanation that relates the late nineteenth-century state of scientific knowledge about hysteria to the elaborate Victorian architecture that was stylish in this era. In his words, by the late nineteenth century, hysteria diagnostic practices "resembled an oversized and slightly vulgar late Victorian edifice— highly articulated in detail and impressive to contemplate from afar, but impractically large and with an extremely shaky etiological foundation."[71] Micale then traces how, in the first couple of decades of the twentieth century, Charcot's elaborate nosology for hysteria was dismantled, and hysteria was absorbed in a variety of new disease categories that were taking shape in this period. These range from epilepsy, which was coming to be understood in increasingly sophisticated terms, to Freud's "anxiety neuroses," which were a reinterpretation of many of the symptoms previously affiliated with hysteria in terms that are recognizable today, such as "anxiety attack."[72]

These historical interpretations of hysteria's decline have also led to conflicting assessments of Charcot's scientific accomplishments. Because hysteria eventually vanished, Charcot's efforts have been called a "scientific failure" from the perspective that the condition that Charcot called hysteria was later recognized as a form of epilepsy that had long been overlooked.[73] Offering a different perspective, Charcot has also been judged as a skilled rhetorician who used the credibility that he had gained through his advances in identifying and understanding other neurological disorders to build a strong group of supporters of his theories of hysteria, even though he was never able to provide the same quantity or quality of scientific evidence for these theories that he had been able to present in relation to other diseases.[74] An even more favorable interpretation is that although hysteria itself vanished as a medical diagnosis, Charcot and his pupils developed some new concepts that were later substantiated in psychiatry and neurology. These concepts include psychiatric theories such as those pertaining to the role of trauma in causing mental disorders, which Freud did not elaborate upon until twenty years after Charcot. They also include insights into the brain's functioning during hypnotism, which could only be based on speculation in Charcot's era but were later documented through functional MRI studies.[75]

Conclusion

Whereas the previous chapter focused on a series of moments when new ideas appeared to emerge, this chapter has focused on a moment when the movement of ideas seemed to stop. Although these kinds of movement have, in rhetorical tradition, been characterized by two different concepts—topos for emergence of new ideas, and stasis for moments of stillness or stoppage in knowledge production—the case of hysteria as analyzed in the previous chapter and this one complicates this distinction between movement and stillness in scientific argumentation. As this chapter's analysis has demonstrated, rhetorical stasis can be compared to the brief pause in a pendulum swing before the pendulum swings back in the other direction, or the pause before the golf player lets loose and swings the club.[76] In other words, stasis is that moment in an argument that is simultaneously full of potential movement and very still. It can be a very loud moment in that many conflicting voices and ideas are making sound, but it is, at the same time, very quiet in that the conflict among these different voices and ideas has become so embattled that the argument has reached a standstill. Serres offers different interpretations of the value of noise in an argument, and his interpretations can be used to explain these apparent contradictions that are inherent in a moment of stasis. As Serres observes, on one hand, noise and conflict can be productive sites of new knowledge, but on the other hand, endless amounts of discussion can sometime be paralyzing. Serres's thoughts regarding the novelty of scientific ideas are also relevant here. He repeatedly emphasizes the need to avoid "the sound and fury of repetitive discussion." He encourages us instead to consider the possibilities that isolation and solitude offer as sources, often overlooked, for new ways of thinking in science.[77]

In terms of new knowledge production, we might say that the noise that was created by the competing ideas in those texts analyzed in the previous chapter was actually very quiet. As we have seen in this chapter's analysis, though, it was the momentum building during that silence that started to spur experts in various new subdisciplines to try pushing their understandings of hysteria to the next stage where the theories could meet the increasing demands for a scientific approach to medicine. Although these experts never truly succeeded (and instead, hysteria itself eventually withered away as a diagnostic category), the new knowledge that these experts produced is worth considering because, as we will see in subsequent chapters, many of the new ways of thinking about the female mind and body persisted long after hysteria vanished as a legitimate diagnostic category. In short, stasis has been revealed in this chapter as both a result and a cause of the types of contradiction that we see in nineteenth-century efforts to understand hysteria.

The nineteenth century was an era of divisions—between science and religion, between the acceptance of ancient beliefs and the desire for new, empirical evidence, between scientific diagnosis and old cures, and between the different disciplines in medicine that emerged. For the female body, this inclination toward division was manifested in an increasing awareness of the body as divided into discrete parts and systems—that is, the brain came to be understood as separate from the uterus, and a clearer understanding of the ovaries' role in female reproduction led to a more precise explanation of sex difference. However, the distinction between women's brains and their genitals was never complete, a point that will become increasingly clear in subsequent chapters. The ultimate effect of this division was an ongoing process of stasis—which began with the accumulation of ancient ideas into the texts analyzed in the previous chapter—then continued as experts in multiple new disciplines tried, unsuccessfully, to dislodge their treatments and new ideas from these ancient beliefs. So the stasis was never completely resolved by this new group of experts. However, they did provide ideas that launched the next phase. We can come to understand stasis as something that does not exist as a discrete moment in scientific history; rather, it unfolds over time and may never be completely finished.

Understanding moments such as these in scientific history is key to recognizing the momentum that causes new knowledge to be produced, and thus, understanding such moments contributes to this book's goal of enhancing our awareness of the many forms of movement that constitute rhetoric. The nineteenth century is interesting and important because over the course of this century, science and medicine eventually were no longer content to sit alongside myths and religious explanations and ancient authorities as equally authoritative sources that possessed the expertise to contain women's troublesome, threatening bodies. Experts—as well as the public audiences of the texts that experts wrote—increasingly expected that science alone would provide the justification for medical therapies, practices, and treatments of conditions like hysteria. Yet at the same time, the experts who sought scientific grounding for their explanations and treatments of hysteria were running into unmovable roadblocks; it is in this sense that the nineteenth century can be characterized as a moment of stasis in scientific arguments about hysteria.

This nineteenth-century era of science is important because the confusion and contradictory beliefs that defined this era served as a rhetorical backdrop from which emerged a rhetorical exigency that made it easy—as we will see in the next chapter—for a concept such as that of hormones to develop and take hold later, in 1905. Building on Serres's notion of topology, examining the points of disagreement or disciplinary noise among the different approaches to the uterus-brain relationship that were current in the nineteenth century illuminates the backdrop or scene against which hormones emerged and that thereby spawned a new understanding of the uterus-brain relationship and sex difference in the early twentieth century. Specifically, as will become clear in subsequent chapters, hormones enabled twentieth-century experts to provide verifiable scientific evidence for theories of sex difference that was based in a brain-uterus relationship that nineteenth-century scientists were never able to provide. Despite the efforts of these nineteenth-century scientists, a disjuncture existed between the kind of evidence that they were able to provide for their theories and the increasing demands for empirical evidence that were made in medicine at this time. In line with Serres's notion of topology, it is this point of disjuncture or noise that produced conditions that were ideal for hormones to emerge as the predominant explanation for sex difference.

4

Stasis Unsettled | The Early Twentieth-Century Rise of Endocrinology

CHICAGO, Sept. 14—A new "chemical tree of life" and the outlines of a new Eden in which man is to enter in the not too distant future, with chemistry acting as the gardener, was presented today before the Century of Progress meeting of the American Chemical Society by four of the world's outstanding chemists.

The new "tree of life" was pictured during a symposium on enzymes, vitamins and hormones, the three recently discovered organic substances without which the processes of life would be impossible. Those participating in the symposium are leading authorities, respectively, on these precious tiny substances.
—William L. Laurence, "New 'Tree of Life' Found by Chemists," 1933

During the same time that Charcot and his peers were trying to sharpen their scientific understanding of hysteria, a diverse group of scientists in other disciplines were investigating a set of chemical phenomena in the human body that would soon come to be known as the endocrine system. The exact manner in which these scientists' activities relate to hysteria is not immediately clear if we focus on the initial context of their research. However, as Serres explains in one of his discussions of topology, sometimes a point that seems far away from another point in terms of physical distance can suddenly become very close. To illustrate this concept, Serres offers the example of a handkerchief: "If you take a handkerchief and spread it out in order to iron it, you can see in it certain fixed distances and proximities. If you sketch a circle in one area, you can mark out nearby points and measure far-off distances. Then take the same handkerchief and crumple it, by putting it in your pocket. Two distant points suddenly are close, even superimposed. If, further, you tear it in certain places, two points that were close can become very distant."[1] Similarly, in the series of events that is narrated in this chapter and the next, we will see arguments emerging in domains that might seem disparate and far away from the discourses on hysteria that were scrutinized in the last two chapters. However, through a series of rhetorical events that occurred in the late nineteenth and early twentieth centuries, these seemingly disparate sets of arguments about hysteria and hormones eventually converged. Understanding this convergence is an important part of understanding how the hysterical woman of ancient times became the hormonal woman of modern times. It also offers additional insights into rhetoric as movement. Specifically, this chapter continues the previous chapter's focus on stasis, but the series of events that is narrated in this chapter highlights the forms of movement that can occur when a stasis begins to unsettle, when the pendulum starts to swing back and its momentum is released. As suggested in Serres's remarks about the folded handkerchief, sometimes this movement can be incited by a collapsing or folding that brings together disciplinary approaches or insights that were previously distinct and far apart.

The chapter's specific focal point is the 1905 lecture in which Starling coined the term *hormone* and the series of rhetorical events, innovations, and discoveries that preceded the coining of that term. In the decades that immediately preceded Starling's lecture, nineteenth-century experts had come to have much faith in the idea of the nervous system as the means through which the body's organs communicated with each other. We saw some evidence of experts' reliance on neurological explanations, for instance, in the previous chapter's examination of late nineteenth-century discourses about hysteria. As we will see in this chapter, that same neurological understanding predominated at this time not only in the context of hysteria but also in relation to emerging explanations of normal bodily functions such as digestion and reproduction. With the discovery of hormones and the ensuing emergence of endocrinology as a discipline, however, the neurological understanding gradually gave way to a more complex model that involved both the nervous system and the endocrine system.

This shift toward adopting chemical explanations of the body's internal communication is a subject that has been critically examined in previous rhetorical and historical scholarship.[2] This chapter extends previous inquiries into the emergence of hormones by examining the rhetorical intricacies of the scientific activity entailed in this shift toward a chemical understanding of the body. After outlining the series of scientific arguments that preceded the 1905 lecture series in which Starling coined the term *hormone*, I consider how this new term and the ensuing growth of endocrinology as a discipline related to the scientific scene more generally at the turn of the century. As I argue, in addition to facilitating a new scientific and popular understanding of the human body, the emergence of hormones also coincided in important ways with the professionalization of science that was underway at the beginning of the twentieth century. Understanding these facets of the discovery of hormones and the rise of endocrinology illuminates the unsettling of the stasis that was explored in the previous chapter and, thus, offers insights into the eventual decline of hysteria as a diagnostic category.

Charles-Edouard Brown-Séquard and the Foundations of Endocrinology

Although there is no record of the use of the term *hormone* prior to Starling's 1905 lecture series, the origins of endocrinology as a discipline are usually traced to 1889, when Charles-Edouard Brown-Séquard—who later came to be known as the "father of endocrinology"[3]—gave a public lecture in Paris in which he proclaimed the miraculous results of his own experiments with self-injected extracts from the testes of guinea pigs and dogs. It is reported that Brown-Séquard's "sensational presentation added spice to a meeting that had gathered to hear routine scientific reports."[4] Thus, although he did not coin the term *hormone* and did not necessarily see himself as an endocrinologist at the time, a rhetorical history of hormones must begin with Brown-Séquard.

Brown-Séquard's scientific work over the course of his career spanned several medical specialties, but the common theme of his interests was his focus on "how one part of the body communicates with other parts."[5] He is credited with discovering "that the action of the central nervous system depends on the networking of functions localized to corresponding areas of the brain."[6] He started out exploring this question from the perspective of neurology, with an interest in the brain and central nervous system, and in 1849, he published a famous treatise on a neurological phenomenon that came to be known as the "Brown-Séquard" effect—a label for his observations about the ways in which specific locations in the brain exercised control over different body parts and functions.[7] He was also known for advances in epilepsy, as he was one of the first to recommend bromides as a treatment for this disease.[8]

In the 1850s, prompted by Thomas Addison's 1855 text on the suprarenal glands, Brown-Séquard became interested in the "internal secretion of glands." Although Claude Bernard is noted as the first physician to use the phrase "internal secretion," in 1855, Addison's monograph and Brown-Séquard's public lectures caused this phrase to circulate widely.[9] Inspired by Addison's writing about the adrenal glands, Brown-Séquard began to conduct his own research about the subject, and in 1856, he published a paper reporting results of experiments in which he removed adrenal glands from animals and concluded from these experiments that "adrenal glands are essential for life" and that the role of these glands was "to remove some toxic factor from the blood."[10] In an 1860 Paris lecture, Brown-Séquard then used the following language to characterize the internal secretions that were increasingly interesting to him and his fellow medical experts in the mid-nineteenth century: "The glands have internal secretions and furnish to the blood useful if not essential principles."[11] This was a novel idea at the time, and it was an important one because it promised to offer the scientific explanation of the body's internal communication systems that experts had long sought. We saw in the previous chapter, for instance, how even Charcot in his effort to provide verifiable scientific explanations of hysteria could only guess when it came to explaining the exact mechanisms through which ovaries, nerves, and the brain interacted with each other.

Brown-Séquard's legacy as the "father of endocrinology" is based on research that he conducted in the last half of the nineteenth century, subsequent to his emerging interest in the body's internal secretions. This research consisted of a series of experiments involving self-injections, the results of which Brown-Séquard reported in his dramatic 1889 presentation in Paris. In the 1889 article in which he reports the results of these experiments, he first expands his speculation about the existence of internal secretions in the seminal fluid: "These facts and many others have led to the generally admitted view that in the seminal fluid, as secreted by the testicles, a substance or several substances exist which, entering the blood by resorption, have a most essential use in giving strength to the nervous system and to other parts."[12] He then makes the more specific suggestion, based on his observation of the behaviors of men between the ages of twenty and thirty-five, that "great dynamogenic power is possessed by some substance or substances which our blood owes to the testicles."[13]

Next, Brown-Séquard describes in great detail the substance that he developed to inject in himself to test his speculations about the regenerative powers of this hypothetical substance in seminal fluid. He describes the substance as "a liquid containing a small quantity of water mixed with the three following parts: first, blood of the testicular veins; secondly, semen; and thirdly, juice extracted from a testicle, crushed immediately after it has been taken from a dog or a guinea-pig." He states that he is seventy-two years old and has made ten injections of this substance over a three-week period. He provides a detailed description to indicate how his advanced age was starting to diminish his energy, causing him to be tired often and detracting from his ability to work long hours in the laboratory. Then, he says that "the day after the first subcutaneous injection, and still more after the two succeeding ones, a radical change took place in me, and I had ample reason to say and to write that I had regained at least all the strength I possessed a good many years ago." He then provides several paragraphs of details about the evidence that he had gathered to substantiate this impression, including measurements of his muscular strength, his urinary strength, and his ability to defecate. He is especially enthusiastic about the positive effect of testosterone on his mental capacities: "With regard to the facility of intellectual labor, which had diminished within the last few years, a return to my previous ordinary condition became quite manifest during and after the first two or three days of my experiments." Then he mentions more specifically the "nervous centres" and the "spinal cord" and indicates that these parts of the body had clearly been impacted by the injections of "spermatic fluid" that he was giving himself.[14]

Despite the initial fervor that surrounded his dramatic public presentations, Brown-Séquard soon after gained a bad reputation among his scientist peers because no one was successful in replicating his findings, and, in fact, it was later discovered that the substances that he had used for the self-injections would have contained only trace amounts of testosterone because of the methods that he had used to produce the extracts.[15] What is interesting from a rhetorical perspective, though, is that Brown-Séquard's dramatic presentation and publication of his results in 1889 not only spawned much commercial activity by others who wanted to sell similar substances in hopes of obtaining similar results but also generated much scientific activity from researchers who wanted to disprove his findings by demonstrating how illegitimate his methods were. It was through these scientists' efforts to dispute Brown-Séquard's claims that "the scientific basis of androgen physiology was laid." These scientists first discovered "the endocrine role of the testes" and then "evidence was accumulated for other chemical messengers."[16] Shortly after Brown-Séquard's 1889 report, a series of other chemical messengers was discovered, and these efforts led to the discovery of other hormonal substances and, eventually, to the "preparation" of insulin in 1922.[17]

An 1893 column in the *British Journal of Medicine* serves as documentation for the rhetorical impact of Brown-Séquard's lecture. As the article reports, "though many jeered at [Brown-Séquard] as the discoverer of the secret of per-petual youth, the notion has steadily gained ground that there is, after all, some-thing in it."[18] This article then offers language that suggests how these ideas had taken shape toward the end of the nineteenth century:

> Physiologists have recently been making a number of observations, which show that many organs do more than what was formerly regarded as their functions. The experiments of Bradford on the kidney have shown that this organ does something else in addition to secreting urine; those of Minkowski and v. Mering on pancreatic diabetes, of Langlois and Abelous on the suprarenal capsules, and of Horsley and others on the thyroid, have led to the introduction of the expression "internal secretion." We think that this term is a rather unfortunately chosen one; but it, nevertheless, expresses that the organs in question have some action on the blood, and through it on the tissues generally, which influ-ences their metabolic changes.[19]

In addition to this brief mention of "internal secretions," this 1893 article delin-eates what remains unknown about the chemical composition of these vaguely defined substances:

> But the precise *modus operandi* is in all these cases still a sealed book. The composition of the internal secretion, where it exists, is unknown. It is, however, presumed that an extract of the fresh organ must contain the active substance in the conglomeration of bodies which are extracted by glycerine, salt solutions, or whatever the solvent used may be. There can be little doubt that these substances are of a complex organic nature, substances which call on the resources of the organism to manufacture for itself.[20]

The article mentions several examples in which organotherapy treatments had been applied. In some cases, these treatments were documented as effective, but in other cases, they were ineffective. In this era, when such therapies were applied—sometimes with success and sometimes not—physicians had no way to determine why they were successful or unsuccessful. It is an interesting side note that hysteria is mentioned in this section of the article as one disease that appears to have seen some success with organotherapy: "Such curative effects are best marked in cases of hysteria and neurasthenia, when there are expectations of relief, and that equally good results were here obtained by inert substances with equal facility."[21] Of course, this observation repeats a centuries-old speculation that hysteria is among those afflictions that are most subject to quack cures because the physical symptoms that the patient manifests are mostly or entirely a product of their own imagination. Although the mention of hysteria is interesting, it distracts from the article's main conclusion, which is that there is something important and novel in the concept of internal secretions, which Brown-Séquard's lectures and research advanced, even if the organotherapy cures that he promoted as applications of his laboratory research were increasingly viewed with suspicion.

Ernest Henry Starling and the Professionalization of Endocrinology

Dr. Ernest Henry Starling's 1905 lectures on "The Chemical Correlation of the Functions of the Body" made a powerful entry into these scientific conversations involving Brown-Séquard and other researchers who were working at this time to learn more about the body's internal systems of communication. Starling's series of four lectures were delivered as part of the Croonian Lectures at the Royal College of Physicians of London on June 20, 22, 27, and 29, 1905. Starling began the first lecture in this series by declaring to his audience that if scientists could come to understand the chemical aspects of this communication system, then they would eventually gain more control over the human body: "If a mutual control, and therefore coordination, of the different functions of the body be largely determined by the production of definite chemical substances in the body, the discovery of the nature of these substances will enable us to interpose at any desired phase in these functions, and so to acquire an absolute control over the workings of the human body. Such a control is the goal of medical science. How far have we progressed towards it? How far are we justified in regarding its attainment as possible?"[22] A few pages into the printed version of the lecture, Starling introduces the word *hormone*. In doing so, he first refers to the Greek word ὁρμάω, or *ormao*, which means "I excite or arouse," and then he explains that these "chemical messengers . . . have to be carried from the organ where they are produced to the organ which they affect, by means of the blood stream, and the continually recurring physiological needs of the organism must determine their repeated production and circulation through the body."[23]

Although this lecture includes the first published use of the term *hormone*, it is reported that the word had originated earlier, following a dinner conversation between Starling and Joseph Needham, who was a biochemist and a Chinese scholar at Cambridge. In discussing the concept over dinner, Starling and Needham had reportedly agreed that a new technical term would be useful, and this realization led them to consult with another colleague, a Classical Greek scholar, who ultimately suggested the Greek word ὁρμάω as a possibility.[24]

In Starling's language, we start to see a shift from past paradigms—in which the brain and central nervous system were understood as the key players in the body's internal communication system—to one in which chemical substances act as messengers to coordinate the body's internal processes. Based on the way that he presents his argument about hormones as a means of chemical control, it seems that he expected his audience to be somewhat resistant to the idea of chemical control. He introduces the notion of chemical messaging slowly and deliberately, leading his audience to accept the idea as a natural complement to, rather than replacement of, the nervous control that was widely accepted at this time. For instance, Starling notes that "in the lowest organisms, the unicellular, such as the bacteria and protozoa . . . the mechanism is almost entirely a chemical one." He then proceeds to observe that "with the appearance of a central nervous system or systems in the higher metazoan, the quick motor reactions determined by this system form the most obvious vital manifestations of the animal." He ultimately asserts that in these more complex creatures, the nervous system had adapted as a means of enabling quick responses, whereas slower processes still relied on the chemical means of control: "Where the reaction is one occupying seconds or fractions of a second the nervous system is of necessity employed. Where the reaction may take minutes, hours, or even days for its accomplishment, the nexus between the organs implicated may be chemical."[25] In this passage, it almost seems as if Starling felt the need to apologize for suggesting that humans may still depend on chemical means of control, which were widely accepted as the primary means of internal communication in the lowest forms of animal life. This existence of chemical control in the human body ties humans to the lower animals, rather than separating them from the lower animals, which would have been problematic because scientists for at least a century prior to Starling had devoted a great deal of effort to distinguishing humans from the lower life-forms.[26]

Later in the lectures, Starling turns his attention to specific bodily processes, addressing how the body's internal communication works and gently introducing the possibility of chemical messaging as a complement to nervous control. So, for example, Starling says the following in regard to digestion: "Further observations showed us that the reaction, instead of being nervous, was in reality a chemical one. The entry of acid into the duodenum or upper part of the small intestine causes the production in the mucous membrane of a chemical substance which we call secretin. Since, as I shall show later, there are other secretins, we may speak of this as the pancreatic secretin. This pancreatic secretin is rapidly absorbed into the blood and travels with the blood to the gland, the cells of which it excites to secrete."[27] In making these claims about digestion, Starling refers to evidence gleaned from a recent experiment in which he and his colleague severed the nerve endings that were previously believed to control the pancreas and observed that pancreatic secretion still occurred in the same way, without nervous control. This was an especially surprising and significant finding at the time because Ivan Pavlov had just won the Nobel Prize for exposing pancreatic secretion as a process that was completely controlled by the nervous system.[28]

In his Nobel Prize acceptance speech, Pavlov described his beliefs about the body's systems and these systems' relations with the external world:

> It is clear to all that the animal organism is a highly complex system consisting of an almost infinite series of parts connected both with one another and, as a total complex, with the surrounding world, with which it is in a state of equilibrium. The equilibrium of this system, as of any other system, is a condition for its existence. And if in certain cases we are unable to disclose the purposeful relations in this system, the reason is that we lack knowledge; it does not mean at all that these relations are absent in the system during its continual existence.[29]

Pavlov then articulates the question that drives his research about digestion, specifically regarding the precise manner in which the equilibrium that he describes in the previous passage is established: "How is this equilibrium effected? Why is it that the glands produce and secrete in the digestive tract the very reagents needed for the successful treatment of the respective object? Apparently, it should be assumed that in some way certain properties of the object act on the gland, evoke in it a specific reaction and cause its specific activity." Pavlov then articulates the question in terms that are even narrower, and he presents his answer to the question in the following passage:

Thus, the purposeful relationship of phenomena is based on the *specificity of the stimuli*, that correspond to similarly specific reactions. But this by no means exhausts the subject. Now the following question should be raised: how does the given property of the object, the given stimulant, reach the glandular tissue itself, its cellular elements? The system of the organism, of its countless parts, is united into a single entity in two ways: by means of a specific tissue which exists solely for the purpose of maintaining interrelations, that is, the nervous tissue, and by means of body fluids bathing all body elements.[30]

In short, we can glean from this passage that Pavlov is confident in his assertion that the body's internal communication processes are controlled exclusively by the nervous system. Some amount of uncertainty about these processes, however, remains in his mention of the second means by which the organism is controlled, which he describes vaguely as "body fluids bathing all body elements."[31]

Interestingly, the remainder of Pavlov's lecture is devoted to summarizing what was known about the first of these two means of control—namely, the nervous system—from both previous research and his own research. He never returns in this lecture to the second means of control, "body fluids bathing all body elements." This uncertainty in Pavlov's language might be said to constitute the boundary at which Starling and his colleagues' research began. In response to the lack of knowledge that is revealed in Pavlov's lecture, Starling and his colleague conducted experiments that led to the discovery of secretin, the digestive hormone that is now widely known as the first hormone to be named and identified. It is likely because of this discovery that Starling was invited to give the Croonian lectures.[32] In 1904, he gave a lecture titled "The Chemical Regulation of the Secretory Process," and in 1905, he gave his four-lecture series. The term *hormone* was first used in the first of the four lectures given in 1905, but then it did not appear again until the fourth lecture, when he used the word seventeen times.[33]

In using language intended to capture the newly emerging understanding of a decentralized communication system within the body, Starling's 1905 lecture both reflected and set the stage for an ongoing debate about the locus of control of human behavior and motivations and for a more nuanced understanding of the body's internal systems of communication. Echoing the arguments of other historians of medicine, Jensen notes that communication was an important metaphor in the early understanding of hormones. Jensen emphasizes that this was not a top-down style of communication that constituted a "rote exchange of orders" but an "active, affective, and intentional" system, with signals moving in multiple directions throughout the body.[34]

One interesting side note to Jensen's interpretation is that in a 1920 publication, Starling's collaborator William Maddock Bayliss expressed some dissatisfaction with the term *hormone*, as he felt that the term did not adequately capture the messaging component that had been an important part of the term *chemical messenger* that was often used to describe these internal secretions before the term *hormone* became available. Nonetheless, Bayliss contends in this publication that messaging was still a defining feature of hormones, whether explicitly present or not in the notion of "setting in motion" that was settled on with the choice of the term *hormone*.[35]

Regardless of these variations in possible interpretations of the word *hormone*, the hormonal understanding that began to emerge in Starling's lecture is a more complex understanding than that which was afforded by a strictly neurologic model of the body's internal communication. In this new model that incorporated chemical messaging, the organs are understood as communicating and interacting directly with each other in processes facilitated by internal secretions that came to be understood in increasingly scientific terms. In this emerging view, neither a single organ nor the central nervous system dominates the communication system.

Although much of the text in Starling's lectures is devoted to nonreproductive bodily processes such as digestion, a lengthy section of his fourth lecture addresses the testes, ovaries, uterus, and mammary glands. We see similar ideas in this language, which starts to unsettle the notion of sexuality and reproduction as processes that were controlled by either the organs themselves or the central nervous system. As a result, Starling's discussion of the role of hormones in the female reproductive system seems to promise a scientific, mechanical, and gender-neutral vocabulary for talking about female reproduction. His emphasis is on laboratory discoveries and interactions among the different organs, and particularly on chemicals that are depicted as operating independently of the woman's brain and central nervous system. His language, in this regard, appears to indicate a sharp contrast to the long tradition of medical texts that were explored in the previous two chapters.

In the fourth lecture, "The Chemical Correlations Involving Growth of Organs," Starling reports what is known at the time about female reproductive hormones. He talks mostly about a number of experiments that had been recently conducted on rabbits, in which researchers attempted to determine which hormones caused specific events during pregnancy, including the growth of uterine tissue and the production of milk. He presents several different theories, including one that posits the ovary as the source of these hormones, and another that posits the corpus luteum as the source. Starling discusses recent research investigating the possible role of hormones in menstruation: "It has been shown recently by Marshall and Jolly, in a paper read before the Royal Society, that the changes in the uterus which determine menstruation are due, not to ovulation, but to an internal secretion arising from the ovary." He alternates between referring to this substance as "internal secretion" and "hormone," as in the following sentence: "These observers suggest that the interstitial cells of the ovary may be the seats of manufacture of this internal secretion or hormone." In this part of the lecture, he also frequently draws parallels between the ovaries and the testes. He says, for instance, "definite evidence is brought forward of the origin of the interstitial cells of the testis from the germinal epithelium and of the complete equipotentiality of those cells with those which are forming the ova and Graafian follicles."[36] The most complete summary statement of what was believed at this time about female hormones is in the following passage:

> These experiments of Fraenckel have been confirmed by Marshall and Jolly, who conclude that the ovary is an organ providing an internal secretion, which is elaborated by the follicular epithelial cells or by the interstitial cells of the stroma. This secretion circulating in the blood induces menstruation and heat. In animals which have been deprived of their ovaries and in which the phenomena of heat are therefore absent, these phenomena can be reinduced by the injections of ovarian extracts. After ovulation the corpus luteum is formed. This organ provides a further secretion, the function of which is essential for the changes taking place during the attachment and development of the embryo in the first stages of pregnancy.[37]

Next, he discusses what is known about the mammary glands, which he characterizes as "a still more striking example of growth in response to chemical stimulation from distant organs."[38] The contested nature of this chemical connection, as he presents it, is parallel to the scientific disagreements that existed in regard to digestion, in that the mammary gland connection was previously thought by many experts to be a neurological connection. As he says, "the nervous nature of the nexus between the generative organs and the mammary glands is still maintained by some writers," but "there are many facts which militate against our acceptance of such a view." He then summarizes the main goal of current research as follows: "Physiologists are therefore ready to believe that the nexus between generative organs and mammary glands is a chemical one, though opinions differ widely as to the seat of formation or origin of the chemical stimulus." Then, after reviewing several recent experiments conducted on rabbits, he articulates a conclusion that suggests current evidence strongly supports a chemical explanation for the communication that had been observed to occur between the mammary glands and other organs in the body: "A consideration of all these results brings us to the conclusion that the specific stimulus which determines the growth of the mammary glands in pregnancy is produced in the product of conception, i.e., the fertilized ovum or foetus."[39]

After a bit more discussion of what is known and unknown about the specific nature and function of "mammary hormones," Starling concludes his lecture, emphasizing the significance of these discoveries about hormones and how they might eventually have a broader impact on medicine:

> The facts which I have been allowed to lay before you will, I trust, serve to convince you of the great part played by chemical processes in the coordination and regulation of the different functions of the body. If, as I am inclined to believe, the majority of the organs of the body are regulated in their growth and activity by chemical mechanisms similar to those I have described, an extended knowledge of the hormones and their modes of action cannot fail to render important service in the attainment of that complete control of the bodily functions which is the goal of medical science.[40]

Returning momentarily to the series of events that was examined in the previous chapter, it is worth noting that this concluding paragraph in Starling's lecture relates in an interesting way to the stasis that was encountered by those experts who worked so hard in the late nineteenth century to achieve a scientific understanding of hysteria. Specifically, this concluding paragraph in Starling's

1905 lecture helps us understand the specific discursive activity around hysteria as a small part of a larger scientific debate that unfolded across the disciplines in the late nineteenth century and that centered around intense efforts to understand, in increasingly precise terms, the means through which different organs in the body communicated with each other. Starling's coining of the term *hormone* was part of a larger trend in medical breakthroughs that soon began to unsettle that stasis. Medical experts at the time were increasingly aware of the elaborate systems and organs within the body and the complexity of the manner in which these systems and organs communicated with each other, but they also faced increasing pressure to offer scientific explanations for these internal communications, rather than the kinds of vague speculations in Charcot's notion of "hysterogenic zones," for instance, or in earlier notions of "animal spirits" or "vapors" that moved through the body.

Stasis Unsettled: Disciplinary and Interdisciplinary Movement

As suggested in the previous chapter's analysis of hysteria, the late nineteenth century was a time when, across the disciplines, ancient views of the body clashed with new discoveries in ways that proved old and new belief systems to be irreconcilable. In the sixteenth and seventeenth centuries, for instance, written descriptions of the human brain[41] and circulatory system,[42] respectively, caused fundamental alterations in understandings of the body that had prevailed since the time of Galen and Hippocrates. However, these new understandings—which were based on evidence that was gleaned from the dissection of human cadavers—had not yet been incorporated into science in a way that caused the old explanations, which were based on animal dissections and speculation, to give way. Thus, for instance, a best-selling text like *Aristotle's Masterpiece* could still, well into the nineteenth century, present personal anecdotes, excerpts from ancient texts, folk remedies, and clinical observations alongside each other.

As shown in Gross, Harmon, and Reidy's comprehensive study of the historical evolution of the scientific article as a genre, scientific articles written in the nineteenth century began to manifest these contradictory epistemologies. The nineteenth-century article, according to Gross, Harmon, and Reidy's study, appears to be a hybrid between old and new ways of presenting science. Thus, for instance, the authors of scientific articles in the nineteenth century still used a considerable amount of personal and nontechnical language, and they still wrote for multiple audiences, both scientific and popular, rather than the more specialized audiences who became typical in the twentieth century.[43] Gross, Harmon, and Reidy describe this style as reflecting a "tension between two views of science, the amateur and the specialist." They go on to note "a gradual shift from description to explanation in many of the sciences, and a consequent increase in the complexity of their supporting arguments" that took place during the course of the nineteenth century, and that "it is this trend toward complexity that will eventually exclude the amateur from serious science."[44]

In tracing the series of rhetorical events that led from Brown-Séquard's dramatic 1889 public presentations on the "internal secretions" to Starling's use of the scientific term *hormone* in the 1905 Croonian lectures delivered to an expert audience at the Royal Society, we can see examples of the larger shift in the genres of scientific reporting that Gross, Harmon, and Reidy's study reveals. For instance, Brown-Séquard's manner of presenting his evidence in large public forums is similar to the rhetorical strategies that Charcot used at approximately the same time to substantiate hysteria's existence as a legitimate medical condition. By presenting large quantities of details that encouraged the audience to visualize the bodily symptoms and the effects of the proclaimed phenomena and treatments, these scientists aimed to convince audiences of the veracity of their claims, but they also, perhaps inadvertently, challenged the audience to go back to their laboratories or clinics and produce their own evidence that might support a different interpretation. This rhetorical style can be seen as fraught with the same contradictions that Gross, Harmon, and Reidy identify as characteristic of scientific rhetoric in the late nineteenth century. The public demonstrations were an ideal venue for such rhetorical work, as they allowed fellow scientists to observe these phenomena and their effects with their own eyes, along with the researcher. In many ways, the style of public lecture delivered by experts such as Charcot and Brown-Séquard can be seen as indicating a transition phase between the prescientific work of earlier centuries and the modern scientific practices that were adopted more widely in the twentieth century, which came to value facts only insofar as they could be incorporated into a larger theoretical explanation.[45] In the absence of consistent technical terms for the many new chemical substances that were discovered in the nineteenth century, Gross, Harmon, and Reidy suggest, scientists had to rely on the "complex phenomenological descriptions" that we see in earlier texts, in which authors devoted a great deal of attention to reporting, for example, the smell, feel, and taste of a new substance.[46] This style was also evident, as we have seen, in Brown-Séquard's written descriptions of his research techniques and in Charcot's written descriptions of his hysteria patients' symptoms and behaviors.

By contrast, in Starling's 1905 lectures, we see a shift toward the rhetorical tendencies that would soon typify twentieth-century science communication. Even though endocrinology did not become established as a discipline until at least a decade after Starling's lectures, these lectures played an important role because the term *hormone* allowed several different scientific principles and emerging ideas about medical treatments to be grouped together. Starling's introduction of a technical term for a concept that had long been vaguely known as "internal secretion" exemplifies, in many ways, these important changes that occurred in the specific context of scientific understandings of the human body, as well as in the broader context of changes that Gross, Harmon, and Reidy's study identifies in the rhetorical practices of scientists in this era. If Brown-Séquard is responsible for drawing public attention to the effects of hormones, Starling is responsible for granting these notions scientific legitimacy. Historian Jean D. Wilson makes an important comment that illuminates how Starling's lecture continued the trajectory that Brown-Séquard began: "In these lectures, Starling summarized the work from his laboratory on secretin and the endocrine control of breast development, introduced the word *hormone* to encompass all chemical mediators, and delineated humoral as distinct from neurogenic control mechanisms. Starling's paper had a profound influence on the field and still constitutes a powerful paradigm, despite the fact that the distinction between the two systems has become progressively blurred."[47] In other words, the new ways of thinking about chemical messaging in the body that Starling presented in his lecture had an important impact on scientific inquiry in the emerging discipline of endocrinology and beyond. Starling's lecture initiated a period of exploding scientific knowledge about this subject. Initially named the Society for the Study of Internal Secretions, the Endocrine Society was founded in 1916, and in 1917, the journal *Endocrinology* was established. But prior to this journal's establishment, a 1916 German textbook cited 2,100 references on the subject. Wilson draws an analogy to the establishment of molecular biology: in both cases, "a new concept provided explanations for and means of investigating many hitherto puzzling clinical and physiological phenomena." Thus, "within 15 yr after 1889 the conceptual framework of endocrinology had been formulated by Starling and . . . within 32 yr endocrinology had evolved from a theoretical concept to a complex, established, and expanding science, an astonishingly rapid development." So, from the perspective of developing new terminology, it seems that Brown-Séquard is responsible for publicizing the term *internal secretions*, which was relatively new at the time when he began his public lectures, in 1889, which Starling then renamed with the more precise and scientifically acceptable term *hormones*. Wilson says that Starling's lecture was important because it "introduced the word *hormone* to encompass all chemical mediators, and delineated humoral as distinct from neurogenic control mechanisms."[48] Having a term was a powerful way to group together what had previously been reported as many individual observations; these observations could not be accepted as scientifically valid until there was a category that grouped them together under one label. This granted earlier observations about internal secretions greater legitimacy because it allowed these observations to be pulled into an emerging theory of bodily communication.

One reason why this new term was important is that, in the years leading up to Starling's lectures, some "organotherapy" treatments showed success and others did not. For those experts who were seeking to develop scientific explanations for the cures that were successful, it was helpful to have a common term that could scientifically denote the substances that were being used in these treatments. Furthermore, the term *hormone* had a scientific legitimacy that "organotherapy" would never have because of its affiliation with quack cures. The concept of hormones, in contrast to that of organotherapy, had emerged from a tradition of laboratory research that was conducted by experts such as Pavlov and Starling who were affiliated with the professionalization of chemistry, as its practitioners sought increasingly to dissociate true chemistry from the ancient practices of alchemy, which they depicted as an amateur pursuit.[49]

In assessing the impact of Brown-Séquard's work, historians of medicine have identified two different branches in these early foundations of endocrinology: a branch that came to be known as *organotherapy*, which was associated with commercial efforts to sell hormonal substances that would supposedly cure various ills without any scientific basis for the claims; and a more scientific branch that quickly evolved into endocrinology. Most historians have placed Brown-Séquard on the side that was accused of being less scientific, and the organotherapy treatments that he recommended were even characterized by many as quackery.[50] He received much publicity for his research, however, which was published in the *Lancet* in 1889, and this research sparked a debate within endocrinology "between cultists and advocates of the scientific method."[51] Some historians of medicine have refuted this origin story, trying to attribute endocrinology to a more respectable origin, but it is now accepted by most historians of medicine that Brown-Séquard's 1889 lecture was responsible for the origin of the field. Wilson's perspective is that even though Brown-Séquard's claims later proved to be false, "the concept of chemical messengers quickly led to breakthroughs that served as the basic formulation of the science."[52]

As Gross, Harmon, and Reidy observe, this expectation of the kind of theoretical explanation that was made possible by the coining of the term *hormone* was an increasingly important element of scientific rhetoric in the early twentieth century. From Brown-Séquard's and his associates' efforts to treat various diseases through what was known as organotherapy—that is, through the attempt to use extracts from various organs in the body to treat disease—the only ones that emerged as useful were thyroid and adrenal extracts. The discovery of insulin in 1922 was what finally gave endocrinology legitimacy, allowing the field to "come into its own."[53]

A 1922 lecture by Francis E. Stewart, "The Growing Importance of Endocrinology and Organotherapy," offers useful insights into this series of events. Stewart first says of medicine in general that "practical medicine as a whole is now undergoing a revolution of a kind calculated to raise it from the 'Art' which it is still termed, to the dignity of a science in which analytical reasoning will supplant empiricism."[54] He then situates endocrinology, or "the science of the internal secretions," as part of this larger transformation in medicine. This author offers two examples—the adrenal gland and the thyroid gland—to illustrate his claim that "the endocrines would introduce an era of precision in practice which would raise clinical medicine to the dignity of a science." In Stewart's discussion of these two examples, it becomes clear that physicians at the time had long been using hormonal substances such as epinephrine to treat conditions such as asthma and pneumonia. They knew these substances were effective, but they did not know why. Thus, much of what these scientists attempted to accomplish in this era was to discern the scientific reasons why these hormonal substances had the effects that they did. They were observing specific bodily effects, such as "increased oxygenation, a rise of the cardiovascular tone, and defensive efficiency."[55]

Stewart highlights endocrinology as a powerful new disciplinary force in this ongoing quest to develop scientific explanations for why these old practices had proven successful for so long: "If many of our past experiences in practical therapeutics are studied from the standpoint of endocrinology, we will discover that in many directions we have been practicing this new branch of medicine without knowing it." Then returning to the topic of specific treatments and conditions, he mentions iodide of potassium, noting that this substance had long been used to treat a number of conditions, without knowing why it worked: "Endocrinology has shown us that when iodides are administered, the thyroid gland absorbs iodine from the salt administered and that it is the thyroid and not the salt which carries on the curative process as a factor of our defensive functions." He then includes a powerful quotation from a physician named Charles E. Sajous: "Endocrinology sacrifices nothing of what we actually know; it adds to and elucidates all we know and, on the whole, it appeals to our highest aims, the relief of human sufferings."[56] Sajous is introduced as being "recognized in India as the most advanced thinker in the Western Hemisphere on all matters pertaining to ductless glands," and at the time, he held the first full chair in applied endocrinology at the "oldest medical school in the United States," the University of Pennsylvania.[57]

In the previous chapter, we saw how science in the nineteenth century was defined by the splintering of medicine into multiple subdisciplines and how experts in these many subdisciplines were reaching a point of stasis or standstill as they became increasingly frustrated in their efforts to understand the internal bodily communication processes involved in hysteria. In Starling's coining of the term *hormone* and the ensuing rise of endocrinology, we can glimpse a form of movement that promised to unsettle this stasis. This turn of events offered hope to researchers in these new subdisciplines who sought to discover the entity that would lead to an all-encompassing explanation of the human body and its communication systems that medicine had sought in recent decades. Interestingly, even though this period of the late nineteenth century was a time of increasing specialization in medicine, with multiple new medical specialties emerging, Stewart's 1922 lecture also places particular emphasis on the importance of the need for different specialists to come together to achieve the best possible scientific understanding of bodily conditions. Along these lines, he quotes Sajous again: "A single branch of medicine was unable to settle any subject of such magnitude, and that this could only be done by the cooperation of all branches of medicine including particularly the clinic."[58] Gross, Harmon, and Reidy's study of the scientific article genre identifies these seemingly contradictory tendencies as another key feature of nineteenth century science: There were increasing divides among the many newly emerging specialty areas, but at the same time, scientists in the new subdisciplines increasingly realized the need for collaboration across the disciplines.[59]

If we turn to one of Sajous's key texts, we can gain further insight into this physician's perception of the need for interdisciplinary cooperation. In this 1922 text, Sajous expresses his opinion that endocrinology itself is at a good place, but he contends that physiology has failed to provide the physiologic explanations that endocrinologists need. In other words, Sajous echoes what Stewart said in his 1922 lecture—namely, that at that time, physicians had a clear understanding, based on empirical observation, that the hormones secreted by glands in the body played an important role in various bodily processes, and that there was still a great deal to be learned about the physiological dimensions of why the hormones functioned in this way. Thus, the ongoing quest to develop scientific explanations for cures that were proving to be successful led to an interesting push and pull among the numerous medical subdisciplines that were emerging and becoming professionalized in the early twentieth century. On one hand, scientists in the newly emerging subdisciplines were claiming to have the knowledge about the newly discovered substances known as hormones that would unlock the mysteries of life, but on the other hand, they were pushing experts in other disciplines to meet these scientists at the boundaries of what they knew, in order that they might bring their knowledge together and come even closer to complete understanding. We might say that they touted their own knowledge while blaming the experts in other disciplines for the shortcomings that scientists in every discipline discovered in their own disciplines' knowledge. Thus, after proclaiming the wonders of endocrinology and its new discoveries, Sajous then emphasizes the importance of cooperation in medicine, which seems to have been an important theme in this era of medicine, even as medicine became increasingly divided into clearly defined specializations. Specifically, he says, physiologists need to work with "clinical medicine and pathology."[60] Sajous's conclusion contains important language that expresses, in general terms, the role that the discovery of hormones played in transforming medical understandings of the human body. He says there that endocrinology is "eminently constructive in the sense that by filling gaps in all directions it finally solves problems of various kinds, which, although near solution for decades, lacked precisely what the endocrines furnish to bring them to fruition."[61] In Sajous's language, we see further evidence of this push and pull between increasing specialization and increasing awareness of the need for crossdisciplinary collaboration.

Thus, the movement that began to unsettle the stasis of nineteenth-century science might be understood as a movement that went back and forth between efforts to separate and efforts to unify the enormous quantities of new knowledge that were discovered in this era. The discovery of hormones, then, can be seen as part of a much larger pattern of movement occurring in the early twentieth century, in which emerging, new subdisciplines of medicine argued for their own all-encompassing explanations, with newly identified substances such as vitamins, enzymes, and hormones being touted by experts in their respective disciplines as the latest big discovery. Having a scientific term to characterize their unique substance gave endocrinologists a legitimate place from which to participate in this new interdisciplinary movement.

Although none of the physicians discussed in this section explicitly mentions any direct relationship to a medical understanding of female problems or even of the reproductive cycle, these physicians' words provide a useful insight into the general state of medicine in the early twentieth century. There was an increased demand for scientific explanations of phenomena such as hysteria that physicians had empirically observed for centuries. Medicine was increasingly being divided into specialties, including endocrinology and many others, as a way to address this demand for more scientific approaches to medicine. It seems that advocates for each of these specialty areas made claims that were similar to those that Sajous made about the ability of their specialty areas to provide these missing pieces of scientific evidence. At the same time, though, experts in the various subdisciplines were also aware of how much they needed knowledge from other subdisciplines.

This back and forth movement between increasing disciplinary specialization and the simultaneous awareness within each discipline of the need to reach across their disciplinary boundaries provides an interesting contrast to the stasis, or rhetorical stoppage, in scientific arguments that confronted experts such as Charcot by the end of the nineteenth century. As noted in the previous chapter, stasis implies stoppage, but it also implies a capacity for movement and change that is similar to the momentary pause that a pendulum makes before moving in a different direction.[62] Recalling these contradictory elements in stasis, if the previous chapter emphasized stoppage, then the present chapter is where we start to see the capacity for movement and change that is contained within that stoppage—a force that would eventually unsettle the stasis that was exposed in the previous chapter.

These multiple scientific events and changes contributed to the situation of stasis that characterized Western science at the end of the nineteenth century. The early twentieth century, by contrast, can be understood as the time when the energy that built up from that temporary stoppage point started to be released, causing the stasis to be unsettled. And just as nineteenth-century efforts to understand hysteria can be taken as indicative of larger scientific endeavors in that era, the unsettling that began at the turn of the century occurred not just at one disease site or in one location but in a widespread fashion across the newly emerging subdisciplines of medicine and across the different geographic locations where science was being conducted.

The Chemical Eden as a New Frontier

In the context of US science, notably, the early twentieth century is the moment when we start to see the frontier metaphor that Leah Ceccarelli traces as a defining feature of scientific rhetoric throughout the twentieth century. As Ceccarelli explains, when this metaphor first emerged, it drew from the image of the frontiersman as a brave explorer who discovered and made use of territory that was previously wild and uncivilized. As that land became increasingly scarce, the same traits were transferred, through the frontier metaphor, to the scientist, with the suggestion that "to be a good American frontiersman in these modern times, one must explore the unknown in remote fields of science."[63]

Reflecting Ceccarelli's observations about the pervasiveness of this metaphor across the scientific disciplines, Leonard G. Rowntree, a Mayo Clinic physician, used the frontier metaphor in a 1925 commentary on endocrinology and organotherapy: "Certain features of the landscape stand out in bold relief. Practically all roads of progress begin in the wilderness as bypaths. It is inspiring to look back over the roads that medicine has traveled and to study the lives of those masters, gifted with divine discontent and a spirit of achievement, who with rude axes blazed the bypaths that led onward to the relief of human suffering."[64] The frontier metaphor provides an interesting counterpart to the notion of movement and unsettling of stasis in scientific arguments. As Ceccarelli notes, scientists' invocation of the frontier metaphor suggests new possibilities and the opening up of new ways of thinking about subjects about which it was previously believed that science had reached its limits.

The emergence of chemical messaging as an additional way to understand the information-processing that occurs within the human body is a key component of this early twentieth-century scientific movement, whether such movement is conceived as the unsettling of a stasis or as the exploration of new scientific territory. Many organs and systems had been discovered and conceptualized a couple of centuries prior, when the dissection of fresh human cadavers was an increasingly a routine activity in medical science. But by the beginning of the twentieth century, the anatomical description of organs that were visible with the naked eye was no longer enough to provide the systematic explanations that scientists sought. Tiny elements such as genes, proteins, and hormones came to be more fully appreciated as the unseen information-processing mechanisms that explained how the body's organs and systems worked together to make the body function. The discoveries of these tiny elements caused the human body to be understood more fully and legitimately as an information-processing system.[65]

The quotation that serves as this chapter's epigraph, drawn from a 1933 article in the science section of the *New York Times*, suggests where hormones were situated in relation to this widespread unsettling. Specifically, in this article's reference to the "chemical tree of life," we see evidence of the fervor that arose from the new "chemical rhetorics" that promised to solve the many mysteries that had confounded scientists since antiquity.[66] By envisioning a new "Eden"— one in which humans will be in control—the above passage also intersects in an interesting way with the frontier metaphor that Ceccarelli documents as pervasive in early twentieth-century US science. If the original Eden was perceived to be sacred because of its pristine, untouched nature, the new Eden that was imagined by the author of this *New York Times* article was more like Ceccarelli's frontier, in that its value resided in the fact that humans were starting to believe that they could cross into it and take control, using its sacred qualities for the betterment of the species. Hormones were among the discoveries around which these new "chemical rhetorics" were built, and as the *New York Times* coverage suggests, hormones played an important role in unsettling the stasis of late nineteenth-century science and helping medical experts at the turn of the century move toward a more nuanced understanding of communication processes within the human body.

As we will see in the next chapter, a great deal of additional rhetorical and scientific movement was required before experts would eventually understand the full implications of hormones as an explanatory concept for specific bodily functions such as human reproduction. That rhetorical and scientific activity will constitute the focus of the next chapter's analysis. As suggested in some of Starling's language in the Croonian lectures, the earliest hormonal discourse of human reproduction seemed to promise a gender-neutral, systematic, and mechanical alternative to the deep gender inequality that had pervaded scientific understanding of male and female difference since the ancient texts of Hippocrates and Galen. Important to note is that Starling's language indicates an effort to treat the male and female sex organs as parallel to each other as he discusses their effects on the body. He says, for instance, "the largest group of correlations between the activity of one organ and the growth of others is formed by those widespread influences exercised by the generative organs on the body as a whole and on parts of the body."[67] He then addresses the effects of removing the testes on the male body and notes that similar effects have been observed in the female body after removing the ovaries. In language such as this, Starling's lectures clearly laid some important groundwork for the new hormone-based understanding of maleness and femaleness that soon emerged, an understanding that seemed to promise more nuanced explanations than the brain-based, uterus-based, or ovary-based understandings that had dominated for centuries prior. As these new explanations developed and were applied to understandings of male and female behavior and societal norms, however, this promise was not necessarily realized.

5

Topology of Sex Difference | A Long History of Men Saying Outrageous Things About Women's Reproductive Organs

One of the most vibrant debates in biology through the ages has been the debate surrounding the question of male and female agency in human reproduction. It was common in the ancient biology texts to treat the female as a passive recipient of the male seed; thus, she was responsible for nurturing, but not creating, the new life. This idea was reflected, for example, in Aristotle's "one-seed" model of conception, which suggested that the male seed provided the form and motive for conception and the female was only necessary because she provided the matter.[1] In the "two-seed" model, by contrast, both male and female were seen as contributing to conception, with each sex producing its own kind of seed. This model was espoused by Hippocrates and Galen. Hippocrates offered as evidence for the two-seed model the fact that both males and females experience pleasure during intercourse.[2]

Although the latter view seems to grant females a more important role in reproduction, it has also led, over the centuries, to the overassigning of agency to females in the case of unwanted pregnancies. In fact, throughout the many decades of history during which ancient ideas about biology remained authoritative, several key medical and legal texts used the so-called two-seed model to deny that rape could result in pregnancy. For instance, Hippocrates asserted that women and men could both expel sperm, so if a woman did not want pregnancy to occur, she could choose to rid herself of the male seed. Galen expressed a related but even more problematic idea that female orgasm was necessary in order for intercourse to result in conception. In these ways of thinking, the female body was in some ways revered for its mystical powers, but it was also seen as more mysterious than the male body in that it was thought to possess a power over reproduction that came from an unknown source. As we have seen in previous chapters, ancient ideas like these continued to have credibility in scientific texts well into the nineteenth century, existing alongside the facts that emerged from new forms of scientific activity such as dissection. More importantly, on the question of rape and reproduction, these ancient ideas about the biology of human reproduction gained added authority because they were instantiated in legal documents. As late as 1814, for instance, Samuel Farr's *Elements of Medical Jurisprudence* echoed Galen's earlier observations about the necessity of female orgasm in conception. Specifically, Farr's legal text claimed that if the woman does not experience "lust or enjoyment" during the sexual act, then "no conception can probably take place."[3]

Of course, from today's perspectives, these ideas seem like quaint and curious relics from a prescientific past. At the same time, however, we also get occasional reminders that these ideas might not be as ancient and forgotten as we would like to believe. For instance, in 2012, Missouri congressman Todd Akin caught the attention of audiences around the world with his public comment that rape was not likely to result in pregnancy because "from what I understand from doctors, that's really rare. If it's a legitimate rape, the female body has ways to try to shut the whole thing down."[4] And Akin is not the first modern public figure who has made remarks like this about the female body. *Washington Post* reporter Sarah Kliff traces a series of comments along similar lines to the 1980s, documenting how such arguments have for several decades been used to deny the necessity of exceptions for rape in antiabortion legislation. Examples that Kliff reports include Stephen Freind's 1988 remark that during rape, "a woman secretes a certain secretion, which has the tendency to kill sperm," and a North Carolina legislator's claim that "the facts show that [for] people who are raped—truly raped—the juices don't flow."[5]

The tendency in recent decades when public figures make remarks like these about the mystical powers that reside in women's bodies has been for news stories to cite medical experts who make reference to scientific facts in order to refute the outlandish remarks. Obviously, so the narrative unfolds in the popular media, no rational person today could ever believe that a woman can use some kind of power that comes from her mind or a substance inside her body to prevent pregnancy when she is raped.[6] This popular narrative is, in many ways, reassuring, as it allows us to believe that we know better now, that today's rational minds (with the exception of a few crackpot politicians who are not seen as scientific experts) have access to better scientific evidence than ever before, so we will never fall victim to outrageous, unscientific myths from the past. As I argue in this chapter, however, this fact-versus-myth narrative of such outrageous public comments about female biology does not tell the whole story. As ridiculous as they seem, the ideas about female biology that were expressed by Akin and his fellow misogynists in the 1980s and 1990s have a basis in the science of sex difference, which assumed a distinct shape in the decades just before the hormonal understanding of sex difference emerged in the early twentieth century. And although these nineteenth-century experts claimed to offer a more scientific understanding of sex difference than their predecessors did, some of their beliefs about women as reproductive beings can be traced to ancient times. Furthermore, as I suggest at the end of this chapter, the earliest hormonal explanations of sexuality and reproduction perpetuated some of these beliefs, even as they sought to provide more mechanistic explanations of male and female sexuality, and even as their authors often claimed that their goal was to benefit women.

This recurring pattern of remembering and forgetting ancient ideas reveals another form of topological movement that permeates the history of scientific rhetorics of sex difference. An examination of such rhetorics as a movement of ideas that is punctuated by moments of remembering and forgetting reveals that hormones became a successful scientific concept not because they represented a new discovery but because they enabled new ways of expressing some old ideas. Specifically, I argue, hormones allowed ancient understandings of femininity to be remembered by allowing scientists to communicate these old ideas in ways that would resonate with audiences who had increasing expectations for scientific evidence to support the beliefs that they were willing to accept as true. Furthermore, the fact that these troubling beliefs about female biology were instantiated in the earliest hormonal explanations of sex difference—explanations that provide the earliest form of what we believe to be true today—also sheds new light on the occasional recurrence of these ideas in public figures' remarks about women today. Whereas the previous two chapters explored moments of stasis or stillness in science and the movement that occurs when a stasis is unsettled, this chapter uses the concept of memory, as elaborated by contemporary rhetorical theorists, to explore the kind of movement that can occur when ideas travel long distances through time and how even, in some cases, ideas that seem long forgotten can unexpectedly resurface again many years later.

Memory, Forgetting, and Scientific Progress

The following remarks made by Serres in *Conversations with Latour* about scientific progress and the passage of time seem especially appropriate as a starting place for a feminist response to the ancient ideas about female reproductive powers that have proven to possess an ability to travel long distances through time:

> Let me say a word on the idea of progress. We conceive of time as an irreversible line, whether interrupted or continuous, of acquisitions and inventions. We go from generalizations to discoveries, leaving behind us a trail of errors finally corrected—like a cloud of ink from a squid.... But, irresistibly, I cannot help thinking that this idea is the equivalent of those ancient diagrams we laugh at today, which place the earth at the center of everything, or our galaxy at the center of the universe, to satisfy our narcissism. Just as in space we situate ourselves at the center, at the navel of things in the universe, so for a time, through progress, we never cease to be at the summit, on the cutting edge, at the state-of-the-art development. It follows that we are always right, for the simple, banal, and naïve reason that we are living in the present moment.[7]

In contemporary rhetorical theory, Serres's concepts find a useful counterpart in Nathan Stormer's explication of memory and forgetting. As Stormer observes, our usual habit is to think of memory and forgetting as being opposed to each other. In this view, memories exist like photographs in a distinct storage place; the images in the photographs can fade with time, but they can also become more vivid, such as when we suddenly remember something that was long forgotten. He says that, in this traditional view, "forgetting selectively chips away at the unity of memory; destroying memory piece by piece and retreating into secrecy."[8] In contrast to this traditional view, Stormer encourages us to think of memory and forgetting as recursively related, and to acknowledge the communal, rather than just individual, dimension of memory:

> For my purposes, the term *memory* refers not to a shared image of the past but to the relations between past and present that are manifested conjointly in ways of life.... Forgetting, although usually treated as antithetical to remembrance, is not the opposite of remembering; it is the condition of

memory.... If memory did not fade there would be no process at all. The past would simply accumulate, never actually becoming "past," but remaining ever present, leaving no need for remembrance. If memory were full and total, if forgetfulness were vanquished, memory would no longer *be* memory.[9]

Stormer then encourages a view of *mnesis* as a capacity, rather than a strategy, which means emphasizing "what particular figurations enable discourse to accomplish, thus highlighting rhetorical possibilities."[10]

With these ideas about time, memory, and forgetting as a backdrop, I argue in this chapter that hormones were a scientifically successful concept because of the new rhetorical possibilities that they created; hormones can be understood as a "figuration" that "enabled discourse to accomplish" new things. In more specific terms, hormones allowed for some old ideas that women were more mysterious or harder to understand than men were to be expressed in a way that made sense to twentieth-century audiences. Thus, I argue, a simultaneous remembering and forgetting is evident in early hormonal discourses that expressed the hope that female bodies could finally be controlled, or could possibly be brought more in line with male bodies, which had always been seen as the ideal. Before, this female mysteriousness was always simply accepted or, in some cases, even feared.

The chapter's analysis unfolds by suggesting some of the ways in which ancient ideas about sex difference "percolated" into scientific theories about sex difference in the late nineteenth and early twentieth centuries, just before and just after hormones started to take hold as the primary means of understanding sex difference. Serres's metaphor of percolation offers a concrete way to conceptualize, in the context of scientific discourse, what Stormer describes as a recursive relationship between memory and forgetting.[11] When we think of time as percolating, rather than flowing, we gain an increased appreciation and awareness of the way in which, at any given moment, it appears that some ideas persist and others get left behind. However, the act of getting left behind is never a permanent situation, as old ideas can trickle down again later, even if they initially got left behind.

The Uterus, the Female Brain, and Their Ever-Changing Interactions

The notion that females are biologically different from males can be traced to the most ancient texts, as we have seen, and for many centuries, this idea was hardly ever questioned, even in the texts of those experts who tried to be the most modern and progressive. The specific language used to articulate biological differences has, of course, taken on many different characteristics throughout history. However, some ideas have been quite persistent, such as the seemingly contradictory beliefs that, on one hand, women are inferior to men but, on the other hand, the female body also possesses some kind of mysterious power that makes it harder to understand than the male body. Previous feminist scholarship offers important insights into the manner in which these ideas can move through time. This movement resembles percolation in that it is highly unpredictable, and the ideas seem to resurface as tiny "drops" that are capable of rejoining in an infinite number of new configurations to produce ideas that seem new but are not necessarily so.

Historian Stephanie Shields characterizes this movement of ideas through time as a process of reshaping, in which each new generation's ideas are molded to fit the needs of the times. Thus, when we take an expansive look at the history of scientific beliefs about sex difference, we can see instances of remembering and forgetting. The language that expresses each new scientific discovery or breakthrough is crafted in a way that enables it to meet the demands of its audiences—whether scientific, public, or both—in a given time period. But the language still corresponds with old, underlying assumptions about biological sex difference. Shields's particular focus is the late nineteenth and early twentieth century, which she refers to as the functionalist era in psychology, in that psychologists' efforts centered on tracing phenomena such as sex difference to particular features or structures in the brain. Shields also traces a shift from functionalism to behaviorism during this period, which occurred partly because of the influence of evolutionary theory. Her analysis emphasizes how, even as this profound shift occurred, the concepts of female inferiority that were established during the functionalist era were adapted to "fit the times" of behaviorism.[12]

Feminist theorist Bernice Hausman also offers important historical insights into the movement that occurred in scientific rhetorics of sex difference in the last half of the nineteenth century, but Hausman's focus is on the ovaries and uterus rather than the female brain. Hausman notes that as the ovaries came to be more clearly understood during this time, these organs took on some of the traits that were previously assigned to the uterus, and for many experts, the ovaries replaced the uterus as the organ that was seen as central in defining femaleness. Hausman also asserts that this shift from uterus to ovaries as the defining feature of femaleness coincided with a shift toward a view of male and female as "incommensurate sexes," which contrasts with the notion of male and female as "two versions of one sex (the dominant paradigm prior to modernity)." Thus, females came to be seen not so much as defective versions of males, but rather, as fundamentally different from, although still inferior to, males. Hausman then traces how the hormonal view of sex difference gradually replaced the ovarian view in the first half of the twentieth century when the field of endocrinology was established. As she argues, "in the shift from ovaries to estrogen as *the* determinant of physiological femininity, we might say that the more things changed, the more they stayed the same."[13]

Extending the feminist historical work of both Shields and Hausman, my analysis in this section suggests that another important aspect of the movement in scientific rhetorics that Shields and Hausman document can be seen in the shifts in discourses about the female brain in the decades just before the emergence of hormones. Whereas Shields focuses on a scientific understanding of the female brain and Hausman focuses on ovaries and uterus, my analysis exposes how ideas about these female organs percolated into late nineteenth-century rhetorics of sex difference and how such percolation enabled a simultaneous remembering and forgetting of ancient ideas about sex difference. As scientific understandings of the human brain increased, experts in various fields offered theories about the unique features of the female brain. These experts sometimes depicted the female brain in a way that suggested that it resembled the uterus or ovaries, and they sometimes suggested that the female brain was connected to the uterus or ovaries in a way that was never indicated for the male brain and sex organs. The specific form of movement that constituted scientific rhetoric of sex difference in this era can be understood as percolation, with ideas from ancient times trickling their way into the newest scientific rhetorics in an uneven and unpredictable fashion that enabled simultaneous acts of remembering and forgetting.

One important development in this regard involved the new types of data that were available as the dissection of corpses by medical researchers became increasingly routine, which enabled these researchers to view the internal structures of human bodies in ways that were becoming increasingly specialized. For the issue of sex difference, having access to this kind of evidence gave experts a new way to support the idea that had always been accepted as true—namely, the idea that females were deficient or inferior versions of males. Rather than theorizing female inferiority as a product of vague concepts such as wandering wombs, humors, energy, or heat, as experts had done for several centuries prior, these scientists believed that they could trace female inferiority to a feature of the female brain. This kind of research led to assertions, for instance, that the frontal lobes in female brains were less developed than those in male brains, or that women's cerebral fiber was softer than that of men. Scientists used these structural differences, which they claimed to have observed in dissected human brains, to account for women's intellectual inferiority and their apparently greater susceptibility to nervous conditions such as hysteria. Other scientists in this time period paid more attention to the total brain size, asserting that women's inferior mental capacity derived from their brains' smaller volume rather than from differences in the shape or form of specific regions in the brain. These arguments assumed many different shapes during the period in question. In this shifting discourse, multiple instances of ideas percolated down through time, with some remembered and some forgotten at any given moment in the discourse.

One important example is an article by George Thomas White Patrick that appeared in 1895 in the journal *Popular Science Monthly*. This article summarized a vast body of interdisciplinary literature about the science of sex differences. Patrick is somewhat ambivalent about the question of brain size and the functions of the brain's different sections. For instance, he says that current literature shows that women had a larger "frontal region" in the cerebrum, which was seen as the most important region for intellectual power, but he then says that contemporary experts place more emphasis on the "parietal region," which was proven to be smaller in women. As he says, "it is now believed, however, that a preponderance of the frontal region does not imply intellectual superiority, as was formerly supposed, but that the parietal region is really the more important."[14] This author, like many others, ultimately acknowledges that the attempt to locate sex difference in the brain had been inconclusive, but as a psychologist, he takes a broader look at bodies and behaviors and thus situates the research about brain anatomy in a wider body of science regarding different kinds of intellectual capacities and other parts of the body. In this author's rhetorical maneuvering, we see a characteristic that recurs frequently throughout the rhetoric of late nineteenth-century science on sex difference—namely, a capacity for movement that always enabled scientists to find a new way to articulate female inferiority, even when existing arguments proved inadequate. In this text, Patrick ultimately ties these different threads together by drawing a comparison between female traits and "infantile traits." When it comes to the brain, he says, "the preponderance of the lower brain centers and the greater relative weight of the whole brain (if the latter be admitted), all these are distinctively infantile marks."[15]

A 1909 article in an anatomy journal reveals a different disciplinary perspective and target audience but reports research goals that are similar to the research that is summarized in the *Popular Science* article. This 1909 article by Franklin P. Mall provides an extensive review and critique of the relevant extant literature. In this article, Mall reports a copious amount of evidence consisting of measurements and observations of brains retrieved from dissected male and female corpses of various ages and races. Each of the various authors whom he cites had tried to identify a feature of the brain that could be used as a definitive explanation that everyone in the research community could agree was true about the observed intellectual disparities among the sexes or races. Because of the contradictory conclusions that turned up in this evidence, in this article, Mall ultimately rejects the idea that sex and race difference can be traced to a specific feature or structure in human brains. Nonetheless, he accepts as unquestioned truth that such difference exists and that white males possess the superior intellect. As he says early in the article, "the smaller frontal lobe in women and in negroes, and the larger in men of genius would prove, it is believed, that this portion of the brain is the chief seat of a good mind."[16] In addition to reporting various studies about the size of frontal lobes, Mall also reports research about other features of the brain. For example, he discusses a body of research that attempts to delineate between simpler and more complex brains. He says, "it is often said that the brains of women are of a simple type," and he then cites studies that involved examining the "gyri and sulci" of different types of brains to determine simplicity and complexity.[17] As with the other areas of research that Mall covers, however, there seems to be no conclusive evidence that these characteristics explain the purported inferiority of the brains of people who are female or nonwhite.

The general approach of Mall's review article is to report findings from studies of a wide range of brain features and structures, then to point out the contradictions or flaws in each body of research that he discusses, and then ultimately to discount this large and diverse body of data on the grounds that the scientists' measurements were not accurate or legitimate. Toward the end of the article, he concludes that "each claim for specific differences fails when carefully tested, and the general claim that the brain of woman type is foetal or of simian type is largely an opinion without any scientific foundation. Until anatomists can point out specific differences which can be weighed or measured, or until they can assort a mixed collection of brains, their assertions regarding male and female types are of no scientific value."[18] Mall essentially concludes that the available evidence was not sufficient to warrant conclusive claims about the anatomical differences among the brains of people of different sexes or races. However, by displaying a capacity for rhetorical movement that is similar to that of other sex-difference scientists in this era, he does not refute his contemporaries' apparent consensus that females and nonwhites have inferior brains. Rather, reflecting the increasing demand for empirical evidence and measurements to support scientific claims, Mall urges scientists to keep looking for "new data" that was "really scientifically treated," so that their claims would not be based on "the older statements."[19] In other words, even though they could not yet find definitive evidence to correlate brain size or features to sex- or race-based differences in intelligence, scientists in this era did not give up on the idea that such evidence would eventually be found, and they remained convinced that their evidence was superior to that possessed by the scientists who came before them. This is typical of nineteenth-century scientific rhetorics of sex difference in that the experts desperately hoped that they could compile evidence that would prove or disprove the ancient explanations that were so long accepted as true. Of course, in the case of sex difference, they were not interested in overturning the assumption that females are inferior; they were instead looking for different forms of evidence that would allow them to remember this long-standing belief but express it in ways that would be compelling to contemporary audiences. Even when they did not find such evidence, they still reported their findings and then used the shortcomings in current evidence as a call for more research.

Another form of remembering that appears throughout these nineteenth-century rhetorics is that, even in scientific discourses that seem to emphasize the female brain, we see examples in which traits that were previously assigned to the uterus were transferred instead to the female brain. In Serres's terms, we might say that these concepts or traits percolated into the brain discourses, in a way that was probably independent of any particular individual's intentions, but these concepts enjoyed a distinct presence there, nonetheless. We start to see this, for instance, in Patrick's 1895 *Popular Science* article, in which Patrick moves beyond the anatomical features of brains and reflects on how the differences between male and female brains are manifested in different mental capacities. He lists a number of intellectual capacities that are superior in women, but he turns each of these into a decidedly negative quality when compared to the intellectual capacities of men. He says, for example, that women have quicker perceptive abilities, that they are better at concrete thinking, have better memory, and are better at "reasoning of the quick associative kind." However, these supposed strengths turn out to be flaws when viewed next to men's intellectual capacities, because whereas men possess "the slow, logical" form of reason that is more important for advanced thought, women's mental tendencies make them more likely to view the world subjectively, rather than objectively: "In apperception the subjective factor is larger in woman, and she sees things more from the standpoint of her own experiences, wishes, and prejudices." And because they think more concretely, women are less able to think about abstract concepts: "For the formation of concepts, especially the more abstract ones, woman's mind is less adapted than man's. She thinks more in terms of the concrete and individual." As for reasoning, women are better at the "quick associative kind . . . but in slow, logical reasoning, whether deductive or inductive, they are markedly deficient. They lack logical feeling, and are less disturbed by inconsistency."[20] He then proceeds to argue that many of these female mental tendencies mean that women resemble not only infants but the "primitive" type. Although his language is ambiguous and he does not necessarily agree that women resemble primitives in all capacities, he points to several features that make women's mental capacities similar to those of savages. After weighing several different arguments for and against this claim that women resemble savages, he concludes that "women's purely intellectual development has been retarded, and this may have a practical significance considering that on these qualities the struggle for existence now so largely turns."[21] This article is interesting because it is written by a psychologist who presumably spent his time observing women's behaviors and actions, not dissecting human brains in a laboratory. Nonetheless, he draws on biology and anatomy research conducted by his peers to try to identify scientific explanations for the behaviors and actions of the females whom he had observed, despite then claiming that men possess a superior kind of reason.

Returning to Serres's ideas of topology, and to Stormer's theorization of remembering and forgetting, in scientific language such as this, we see how the emphasis on the deficient female brain did not completely replace the earlier idea of the uterus or the ovaries as the seat of femininity. Rather, it is more accurate to say that the female brain remained connected to the uterus and ovaries, or became conceptually grafted to the sex organs in a new way, and these connections were disguised in terminology that sounded more scientific. In texts such as these, we glimpse the manner in which hysteria would eventually transition to hormones. In many of these texts, we see scientific authors eliminating any explicit mention of the wandering womb; even when they address hysteria, they treat it as a brain disease rather than a uterus-based disease. Thus, it might appear that the authors have made a complete separation between the uterus and the brain, but they have actually, in many ways, replaced the wandering womb with a uniquely female brain. This brain replaces the womb that used to serve as the explanation for various female behaviors, and ironically, this female brain mimics many of the same behaviors that were attributed to the wandering womb. Women's intelligence was depicted as "cunning," for instance, which is a trait that we might typically assign to animals in the wild, whereas men were said to exhibit the high-level forms of reasoning that were presumably necessary to succeed in the modern world.

We might say that in ancient medicine, when most female problems were attributed to the wandering womb, the uterus had a much more important role to play than the brain did. But as scientific interest in the brain grew throughout the nineteenth century, these forgotten ideas about the uterus (or, more recently, the ovaries) were remembered in new formulations that ascribed the old, forgotten traits of these sex organs to the newly understood female brain. We also see a transition from focusing on the traits of the female brain to focusing on how these traits manifested in female behaviors, which indicates a trend that became more pronounced in the early twentieth century. Shields's historical analysis suggests that, with a shift toward behaviorism, as opposed to functionalism, the scientific quest to explain female inferiority took on another form, as some experts placed less emphasis on the structure of male and female brains and more emphasis on the ways in which these different structures resulted in different observable behaviors. So, for women, it was posited that more "primitive" parts of human nature were expressed in their personalities. Women were more instinctual and more emotional, and these traits were offered as an explanation for their weakness and perpetual failure.

In a 1901 article titled "The Physiological Mental Weakness of Women," German neuroscientist Paul Möbius tried to put a positive spin on this inequity by arguing that women's inferior intellectual ability was necessary for the survival of the human race. As a neuroscientist, Möbius approached the problem as one of classification—in other words, by determining which mental capacities should be expected in different types of humans and then judging individuals' mental capacities in relation to those group-specific expectations. Even though some of his language suggests a desire to create classifications that treat women and nonwhite males as different but equal, Möbius ultimately adopts a negative stance toward women's mental capacities, describing women, for instance, as "dead weight" on men. Women were somewhat necessary but at the same time quite problematic: "She prevents much restlessness and meddlesome inquisitiveness, but she also restrains him from noble actions, for she is unable to distinguish good from evil." The hierarchy in the different mental capacities that he specifies for various groups is made especially clear in the following passage: "The normal condition of the child is pathological in the adult, that of woman in man, that of the Negro in the European. Comparison of various groups is then the salient point, for only thus can it be learned what is to be expected of a member of a certain group, only thus will one avoid calling a person stupid or mentally weak, because he does not do what some other person is capable of." Möbius then poses the question that seemed to be on so many scientists' minds in this period: were women's mental faculties just different than men's, or were they inferior? In response, he says, "a sea of ink has been wasted in this matter and still there is no harmony and clearness." He then summarizes contemporary disagreements surrounding questions such as whether absolute or relative brain size was more important and whether brain size was even an accurate predictor of mental capacity. To a greater extent than some of the other scientists in this period, however, Möbius seems willing to accept that scientists will eventually identify some specific feature in women's brains—whether total size, relative size, or size of a particular region in the brain—that explains her intellectual inferiority. For instance, the following statement is italicized, seemingly indicating that Möbius accepts this summary statement to be true: "It is therefore demonstrated that *parts of the brain extremely important for mentality, the frontal and parietal convolutions, are less developed in woman than in man and that this difference exists at birth.*" Not surprisingly, Möbius concludes that sex difference was defined by different mental faculties, with women's—in contrast to men's— mental faculties being governed more by instincts: "One of the most essential differences is that instinct in woman plays a greater role than in man."[22]

In this language, we see a particular form of remembering and forgetting of ancient ideas about the uterus that becomes even more important in subsequent decades. Rather than assigning the female brain characteristics that were previously assigned to the uterus, with language such as Möbius's, we see the uterus and female brain depicted as having existed in a specific relationship, with women's reproductive capacities having exercised an influence on the brain that was not postulated for male reproductive organs and brains. Along these lines, Möbius opposes instinct and "reflection," and of course, he says that men's actions were motivated by reflection, whereas women's were motivated by instinct. He later speaks of "feminine cunning," which seemed similar but, as he says, such cunning "is not a sign of high mental endowment; woman in this respect is as inferior to man as a shrewd merchant is to an artist or a scholar."[23]

Möbius also suggests that women were "instinctively adept" and that this was "an essential part of the feminine nature." He then makes one specific mention of the uterus, using language that reinforces the idea that the uterus was more than an organ inside the woman's body. Rather, the uterus was fundamentally connected to the brain in women and, thus, contributed to having made their mental powers inferior to men's: "Just as an intelligent man would not seek a literary woman to care for his children, eternal wisdom has not placed beside man a man with a uterus, but the woman to whom all is given requisite for a noble calling, but to whom is also denied masculine mental power." Furthermore, he then says that the female brain was a necessity because "if we wish a woman to fulfill all her maternal duties she cannot have a masculine brain."[24]

Next, Möbius explains some of women's specific biological functions, such as menstruation. Some of this language can be seen as a precursor to the ideas that, after the discovery of hormones, were more deeply engrained in the scientific understanding of women's inferiority. He says, "woman during a considerable portion of her life is to be regarded as abnormal. I need not speak to physicians of the bearing menstruation and pregnancy have on mentality, to show that both states, without actual disease, disturb the mental equilibrium, impair the freedom of the will within the meaning of the law."[25] In short, it seems that old ideas about the uterus were both forgotten and remembered in late nineteenth-century and early twentieth-century discourses about the female brain, as experts increasingly assigned traits to the female brain that were previously assigned to the uterus. We might even say that, through this process of remembering and forgetting, the female brain in scientific thinking gradually evolved between the seventeenth and early twentieth centuries to take on the characteristics that were previously affiliated with the womb.

According to Shields's historical analysis of the science of female brains, the ideas that we see in Möbius's text gained prominence in the mid-twentieth century, as scientific attention to maternal instinct gained intensity. Although it was widely accepted at this time that humans were separated from animals by their greater reliance on reason and intellect rather than instinct, the maternal instinct was seen as an exception to this rule and, of course, an exception that applied only to women.[26] Shields then demonstrates that in the mid-twentieth century, maternal instinct came to be understood more in terms of "drive" and "motivation" as it was integrated with behaviorist ideas; this development led scientists in this era to be increasingly interested in studying maternal behaviors in rats and other rodents. The notion of maternal instinct also took on more negative connotations at this time, such as being affiliated with masochism and sexuality and, thus, came to be seen as potentially destructive.[27]

Interestingly, this notion of instincts as an explanation for female behavior persisted well past the 1905 emergence of the term *hormone*. For instance, a 1919 article by Knight Dunlap titled "Are There Any Instincts?" defines instincts as "any responses that have not been learned." He then proceeds to outline what appears to be a persistent tension in psychology between defining instincts as "a group of activities teleologically defined" and "the instinct as a physiological group."[28] Although Dunlap's article was written well after the term *hormone* was coined, he does not use the term in his article. Other than one instance of the word *glandular*, there does not seem to be much connection to hormones in this article; as a psychologist, the author seems more interested in behaviors than in the physiological causes of those behaviors. Taken together with other texts, Dunlap's article suggests that instinct was a controversial subject at the time. There was no consensus about how to define it, and there was not even clear agreement that a "maternal instinct" was universal among women. For instance, a 1914 article by Leta Stetter Hollingworth, questions whether anything resembling maternal instinct could be strong enough to compel women to endure the extreme pain of childbirth: "There is no verifiable evidence to show that a maternal instinct exists in women of such all-consuming strength and fervor as to impel them voluntarily to seek the pain, danger, and exacting labor involved in maintaining a high birth rate."[29] This article is important because Hollingworth points to social circumstances that led women to become mothers and to devote their energies to child-rearing; she tries to downplay the importance of something like instinct, or some other biological capacity, that made women more inclined toward mothering and domestic activities. The concept of instinct never gained much scientific credibility, but it is important because it served as an early attempt to convey an understanding of the uterus-brain relationship that held up to scientific scrutiny. The notion of maternal instinct, as described in these early twentieth-century texts, rested on the assumption that something inside a woman's body controlled her brain—that is, that she could never have been an entirely free-thinking, rational individual in the way that men were assumed to have been.

Alongside these ideas about yoking women's brains to their uteruses, we see a remembering and forgetting of ancient ideas about femininity in the late nineteenth and early twentieth centuries in arguments that developed around the notion of an opposition between the uterus and the brain. Thus, for instance, experts expressed serious concerns about the possibility that if women engaged in too much intellectual activity, their reproductive capacities would be harmed. Möbius, for instance, discusses the dangers inherent in women's intellectual activities and posits an opposition between the brain and the reproductive organs: "Evidently the primary phenomenon is the opposition between brain activity and propagation. Both functions are closely united, but more one gains the ascendancy the more the other suffers." This text also includes a lengthy discussion of the potential dangers that would have arisen if women had tried to become more engaged in intellectual pursuits such as medicine. His language about this subject includes admonishments about the possibility that women's breeding and child-rearing capacities would be diminished if they tried to thwart nature's intentions: "Nature is a strict mistress and threatens rigorous punishment for the violation of her commands."[30] Along similar lines, nineteenth-century scientist Dr. Edward Clarke believed that if too much blood flowed to a woman's brain, she would not have had enough blood to support the reproductive organs and thus she would not have been able to fulfill her reproductive capacity if she engaged in too many too-rigorous intellectual pursuits. A different version of this concern was expressed by some scientists who worried that if women and men were educated together, they would be in competition with each other, and such competition would threaten the future of the human race by leading to less reproduction.[31]

Understanding these many different configurations of the female brain, uterus, and ovaries and the connections among them is another way to illuminate the disciplinary divides that, as discussed in earlier chapters, occurred frequently in the nineteenth century. As we have seen, the period from the late nineteenth century to the early twentieth century was an especially important historical period, as it was the time when many new medical specialties emerged and began to define themselves. On one side of these disciplinary divides were the biologists and anatomists who focused on human brains as the main motivator of human behavior and, therefore, sought to explain women's inferior intelligence by discovering some aspect of female brains that differed from male brains. On the other side were the psychologists, psychiatrists, and neurologists who focused on such phenomena as behaviors, feelings, and instincts and, as a result, paid much more attention to women's mental health and behaviors than to their intelligence levels. The net effect of these different scientific quests across the disciplines was a multifaceted form of movement in scientific rhetorics that can be characterized as a simultaneous remembering and forgetting of ancient ideas about masculinity, femininity, and the differences between them.

Emergence of Hormonal Theories of Sex Difference

When the term *hormone* was coined in 1905, it seemed initially that scientists might get distracted from the long quest to explain male-female difference through structural or functional features of the sex organs or the brain. As suggested in the previous chapter, Starling's language about reproductive hormones reflected a more neutral, mechanical view than did the misogynistic language that had been used in scientific explanations of sex difference for so many centuries prior. As I argue in this section, however, the earliest hormonal explanations of sex difference started, ever so subtly, to do the rhetorical work that would later prove necessary to ensure that hormonal explanations of sex difference would not erase the assumptions of female inferiority that had infused the science of sex difference since ancient times.

On topics such as sex difference, which had for so long been surrounded by myths and spiritual explanations, the new "chemical rhetorics" that began to take hold in the early twentieth century promised to open up more neutral, mechanical ways of thinking.[32] We see this promise in many of the textual examples considered in this section. These key texts enable us to explore how hormonal explanations replaced the neurological explanations of human reproduction that immediately preceded them. We will also see in this section, however, that even in these early, seemingly neutral explanations, some early traces of the gender inequities in hormonal discourses will become more pronounced from the mid-twentieth century onward. Three interrelated themes emerge in the early research about sex hormones that is discussed in this section. First, since the beginning, researchers have paid far greater attention to female sex hormones than to male sex hormones. Second, the overarching goal of research about female sex hormones, in contrast to similar research about male sex hormones, was to understand and chemically synthesize these hormones so that women could be controlled through scientific and technical manipulation. Third, some of the research discussed in this section approaches the female body from the assumption that femaleness was something to be cured or corrected, but never enhanced, as seemed to have been the case with the relatively small amount of research about male sex hormones that was conducted in this era. These related ideas can be understood as a simultaneous forgetting and remembering of old ideas about the female body as an entity that was hard to control or dangerous. That is, many of the particulars of these old beliefs seemed to disappear from the earliest instances of hormonal discourse, but at the same time, some of the old ideas were remembered in a hormonal discourse that tried to communicate femaleness in a more scientific and less disparaging way.

As had been the case for other bodily processes such as digestion, incorporating hormones—and the related network theory of bodily communication that ideas about hormones were used to support—led to a shift in theories of human reproduction that moved away from an exclusively nerve-based understanding to something more complex. Historical analysis suggests, however, that scientific understandings of reproduction proved more recalcitrant to hormonal explanations than did other bodily functions.[33] This was partly because the effects of sex hormones were harder to trace than the effects of earlier hormones were. Scientists and clinicians could measure the effects of adrenaline, for instance, by documenting changes such as increased blood pressure. By contrast, in the case of sex hormones, researchers were studying a complicated set of interactions that involved both male and female biological components, and except for the visible effects that occur in the female body through processes such as menstruation, pregnancy, and lactation, these interactions and bodily effects were largely invisible and impossible to measure.

Despite these early difficulties, scientists did eventually come around to the hormonal view of reproduction, and as had been the case for previous bodily functions, Starling's introduction of the hormone concept served as a catalyst in the efforts to solve many of the puzzles that scientists had long been trying to solve. Historian Merriley Borell postulates that hormones were an important concept in the context of sexual reproduction because they offered a "less controversial" means of understanding the "theory of internal secretions" that researchers had previously struggled to make sense of in the context of reproduction. This breakthrough was also possible, according to Borell, because researchers adopted new strategies. Instead of looking just for measurable effects like changes in blood pressure, they started to do more work that involved physical manipulation of animal specimens, such as removing ovarian tissue, using grafting techniques to replace it, and then observing the impact on various organs and functions in the animals' bodies. Borell also identifies important moral and social dimensions to this early twentieth-century research about sexual reproduction. As he and many other historians have noted, these topics were previously considered off-limits for scientific study, but moral attitudes toward sexuality became less rigid in the early twentieth century. Furthermore, as Borell notes, hope increased that scientific study of these previously forbidden topics would contribute to solving many of the problems that human sexuality had previously been perceived as causing. In Borell's words, "traditional resistance to scientific study of the reproductive process was overcome in the early twentieth century by the expectation that this particular line of inquiry would lead to understanding and therefore medical control of the complex of disorders—physiological, psychological, and social—associated with the generative glands."[34] It is worth noting that, of course, throughout history, the disorders that Borell enumerates were associated with the female body, not the male body. This offers one hint of a possible reason why more attention was paid to female hormones than to male hormones since the beginning of recorded history.

These various breakthroughs contributed to what Jensen calls the "rise of reproductive endocrinology," which, according to her analysis, occurred between 1926 and 1940. By 1910, according to Jensen, experts "had garnered enough evidence on the topic that the vast majority of scientists and medical practitioners had accepted the idea of the existence of testicular and ovarian hormones and their function as chemical catalysts and bodily regulators."[35] The rhetorical shift that Jensen identifies can be traced to hormonal explanations of reproduction that emerged soon after Starling coined the term *hormone*, when scientists started to apply the concept more specifically to the understanding of male-female difference and of reproduction.

As emphasized in the previous chapter, Starling's introduction of hormones as a way to explain the body's internal communication seems to promise a gender-neutral, mechanical way to explain such concepts as sexuality, reproduction, and sex difference. One important difference to note in Starling's treatment of male and female bodies, however, is the amount of text that is devoted to male and female sex organs. He devotes one paragraph to the secretions from the testes that are believed to be responsible for male hormonal effects, but he devotes the remainder of his lecture to female organs and hormones. In this section of the lecture, Starling addresses the mammary glands, uterus, and ovaries, and he offers an extensive discussion of experiments that were conducted on pregnant, nonpregnant, and virgin rabbits to try and understand these different hormones and organs and how they functioned in the body. This disparity in the amount of attention that Starling pays to male and female hormones is part of a trend that continued into the early part of the twentieth century. Another possible explanation for this disparity is that because menstruation, pregnancy, and lactation in the female body were the most visible signs of the human reproductive cycle, these were the bodily processes that also received the most attention from researchers who studied hormones.

The suggestion that the ovaries had something to do with menstruation was documented as early as 1793, but this idea did not become part of general medical knowledge until the 1860s.[36] Scientific understanding of the ovaries' involvement in menstruation made its next advance when Ludwig Fraenkel reported studies on rabbits in 1903 and 1910, demonstrating that the corpus luteum prepared the uterus for the implantation of the embryo. The challenge of understanding the relationship between the estrous cycle in nonhuman mammals and the menstrual cycle in women remained an important stumbling block at this time, however. For a long time, researchers had assumed that the period of "heat" in nonhuman mammals was comparable to menstruation in women.[37] This explanation seemed to make sense because these bodily events were both marked by bleeding. Two key studies, published in 1908 and 1911, respectively, proved these assumptions to be erroneous, so that by the second decade of the twentieth century, expert consensus had shifted toward our current understanding that for humans, ovulation occurs around the middle of the cycle and menstruation at the end. As Corner reports, "with this information, the human cycle becomes a clearly understandable special case of the generalised oestrous cycle, from which it differs in two major points—namely, that sexual responsiveness is not limited to a sharply defined period at the time of ovulation, and that the corpus-luteum phase terminates more stormily than in lower mammals, with haemorrhage from the endometrium." Even after these important breakthroughs, however, Corner reports that much work was still needed to understand exactly how the corpus luteum "acts to set up the menstrual flow."[38]

A 1914 study by F. H. A. Marshall and J. G. Runciman was an important next step in this direction. This article appears to be one of the first texts to use the word *hormone* in relation to the "chemical substances" that governed the female reproductive system. Even though Corner reports that the human menstrual cycle was separated from the animal estrous cycle a few years prior to this study, Marshall and Runciman still conflate the estrous cycle of animals with the menstrual cycle of humans. The authors begin by summarizing recent experiments to determine how the surgical removal of ovarian tissue impacts the female menstrual cycle. They then claim that "the inference drawn from these experiments is that heat and menstruation are brought about by some chemical substances or hormones which are elaborated by the ovaries and act either directly or indirectly upon the tissues of the uterus and mammary glands, as well as upon all the other organs which undergo cyclical changes associated with the female sexual functions." The authors then describe the goal of the present study as determining "what part of the ovary" was responsible for the female menstrual cycle. They dismiss as incorrect earlier scientists' understandings, which "assumed that the functional correlation between the ovaries and uterus was nervous in character."[39] The authors report the results of a series of experiments that was conducted on dogs. In these experiments, scientists had conducted surgery to examine the follicles in the ovaries prior to estrus, and then they killed each dog after estrus so they could examine its uterus and ovaries. Their conclusions pointed the way toward understanding hormones as the impetus for the estrous cycle, but the authors are not precise in the language they use. Specifically, they conclude that "the view which has generally been maintained, that the ripening of the Graafian follicles and the onset of menstruation or heat stand to one another in relation of cause to effect, must be finally abandoned. It is probable that both series of changes are effects of some more deep-seated ovarian phenomenon."[40] In this vague reference to an "ovarian phenomenon," the authors presumably refer to what would later become understood more specifically in terms of the types of hormones that could be specifically identified as the impetus for the different phases of the female reproductive cycle.

A 1927 article by Chauncey N. Allen, "Studies in Sex Differences," illustrates how the gender disparity in hormone research became more pronounced in the early decades of the twentieth century. Allen's article begins by discussing sex differences in relation to skull size, and he references some of the same ideas about brain size and bodily features that were discussed earlier in this chapter. In particular, Allen emphasizes research that indicated differences in rates of maturation for males and females; he suggests that females matured more quickly, but female maturation then slowed after pubescence. But later in the article, Allen shifts focus and discusses hormones and the endocrine system. As he does so, he does not entirely leave behind the behavioral matters that we have seen in the previous organ-based explanations of sex difference. In fact, he moves quite easily from the chemical effects of hormones on the organs to the "problematical effects" that these hormones had on "female behavior" because of their role in menstruation and ovulation. As Allen says, "it is necessary to con- sider the real and problematical effects of endocrine secretions, especially those from the gonads, and the influence of menstruation and ovulation upon female behavior." Allen then discusses "studies on animals" that showed "that the phe- nomenon of heat and the ripening of follicles, for example, are probably effects of some deeper changes in the ovaries." He cites the 1914 Marshall and Runci- man study to support this speculative assertion: "This is another indication of the trend towards explanation of maleness and femaleness in terms of hor- mones."[41] Because this text contains a mixture of many of the different ideas that we have seen in earlier texts, it offers an especially powerful example that illustrates the remembering and forgetting of ancient ideas about the female body that characterized the hormonal view of sex difference that emerged in the first half of the twentieth century.

A 1929 article by George W. Corner and Willard M. Allen is documented as the first scientific text that explicitly designates hormones as the impetus for menstruation. This study, in contrast to Marshall and Runciman's 1914 study, clearly acknowledges the distinction between the estrous cycle in animals and the menstrual cycle in humans. Specifically, the authors say that "it seems prob- able . . . from our experiments that the progestational proliferation is caused by a specific hormone, differing from oestrin, which is elaborated by the corpus luteum; and this is our present working hypothesis."[42] So, as these authors sug- gest, the female hormone oestrin (later known as estrogen) had already been discovered and was starting to be understood; this was first reported in a 1923 article by Edgar Allen and Edward A. Doisy.[43] But this Corner and Allen study, published six years after Allen and Doisy's article, was an early attempt to understand a second female sex hormone, which in 1934 would come to be known as progesterone. This finding is stated more explicitly and definitively at the end of Corner and Allen's article: "It appears, therefore, that the extracts of corpus luteum contain a special hormone which has for one of its functions the preparation of the uterus for reception of the embryos by inducing progesta- tional proliferation of the endometrium."[44]

This attention to female hormones can be seen as a new version of the ancient belief that female biology is more complicated than male biology is; the new element that was introduced in this expression of the ancient belief was a newly emerging hope that advanced scientific knowledge could eventually free women from the aspects of their biology that had long been understood as troublesome. This expression of hope is evident, for instance, in a 1929 book, *The Female Sex Hormone*, by Robert T. Frank. In this book, Frank describes the goal of his contemporaries' quest to discover and chemically synthesize female sex hormones: "Since 1923 the subject has attracted innumerable workers who are elbowing and jostling each other and jockeying for position in the neck and neck race to isolate and synthesize the much desired and long sought for hormone, which is bound to relieve many of the ills from which women now suffer."[45] So again, this passage in Frank's book substantiates my argument in this chapter: An important element of the transition from organ-based to hormonal understandings of sex difference was a remembering of the ancient belief that women's biology was harder to understand than men's, and a simultaneous forgetting of the idea that this biology was inescapable. Rather, in this early twentieth-century version of these beliefs about sex difference, it was suggested that women might eventually find relief from the condition of being feminine through the discovery of the chemical substance (i.e., hormones) that made them that way and the eventual application of the knowledge that would result from this discovery.

The emphasis on using hormonal discoveries to free women from their problems contrasts importantly with the language that was used to justify early research about the male sex hormones; the research about female sex hormones seemed to share the goal of making women less feminine (e.g., getting rid of "female problems"), whereas the research about male hormones sought to make men more masculine. For instance, a 1923 article by Calvin P. Stone is a relatively rare example of a research report that displays a focus specifically on male hormones. Stone continues this discussion of the shift from a neurological to a chemical explanation of reproduction, claiming that scientists turned to hormonal explanations because they were unsuccessful in providing neurophysiologic explanations of sexual behavior. The particular focus of Stone's article is testicular hormones, and the author suggests that there was not much knowledge at the time about this subject: "Factual data concerning the mode and seat of action of the testicular hormone are wanting. At the present time, however, it is generally believed that its influence is exercised by direct action upon the central and sympathetic nervous system and through chemical regulation of the general metabolism."[46] Stone reports that the therapeutic effects that were observed through application of the testicular hormone were positive, including such benefits as rejuvenation and increased sexual appetite. Of course, this stood in sharp contrast to the promise of research about female hormones—

namely, that the knowledge and technological applications resulting from this research might have eventually freed women from the problematic effects of their hormones.

Conclusion

This chapter's analysis has traced how a set of ideas about female biology were fundamentally altered through interconnected processes of remembering and forgetting in the decades immediately before and after Starling coined the word *hormone* in 1905. The set of ideas that are communicated in these late nineteenth- and early twentieth-century texts, I have argued, enact both a remembering and a forgetting of ideas that had been expressed several centuries earlier in ancient Greek texts. In particular, we see in these late nineteenth- and early twentieth-century texts many different versions of the ancient idea that female sex organs were wild animals, that they were inherently harder to control than male sex organs were, that they were harder to understand than male sex organs were, and that femaleness was generally surrounded by a degree of darkness and mystery that was not present in the male body. In reflecting on Stormer's theorization of memory and forgetting as actions that are fundamentally implicated in each other, I am not arguing that the modern scientific rhetoric that communicates new versions of these ancient ideas simply produced copies or reproductions of authentic originals from Ancient Greece. Rather, I have suggested in this chapter's analysis that these ancient ideas were first moved—through an interconnected process of remembering and forgetting—from uterus to ovaries to female brain. Then, as hormones were integrated into the science of human reproduction, these same ideas came to be remembered and forgotten again as scientific explanations of the uterus-ovary-brain connection morphed into the new configuration that soon became the hormonal woman. On one hand, through the series of events that were reported in this chapter, we see a remembering and forgetting of the idea that the uterus controlled women's behaviors at the same time that traits that were previously assigned to the uterus came to be assigned to the female brain. On the other hand, we also see a remembering and forgetting that surrounded the very notion of organ-based sex difference, as organ-based theories began to give way to the more precise understanding of bodily communication that was afforded by increased understanding of hormones as chemical messengers. In early versions of this emerging hormonal model of femaleness, we see how a more scientifically credible version of the uterus-brain connection became possible as scientists came closer to a more precise understanding of female bodily processes, such as menstruation, that would eventually be afforded to them by hormones.

As experts' focus shifted from the uterus to the female brain, old ideas about the uterus percolated into the female brain discourses, and traits that were previously assigned to the uterus were assigned to the female brain. Thus, the female brain gradually evolved in scientific thinking between the seventeenth century and early twentieth century to take on the characteristics that were previously affiliated with the womb. Next, scientific attention turned to the ovaries. Shortly thereafter, as we will see in subsequent chapters, a series of rhetorical events led the general concept of hormones to be shaped more specifically into types of hormones such as sex hormones and, even more specifically, into female and male hormones. Through close analysis of scientific texts that were published in the prehormonal discourses of the several decades just prior to the 1905 emergence of hormones, this chapter has exposed how scientists in this prehormonal era sought to achieve increasingly precise scientific explanations of the uterus, the ovaries, the female brain, or new concepts such as the maternal instinct as the dominant forces that governed women's behaviors. If we want to comprehend the science of sex difference as it has developed over the last few centuries, then it is necessary to examine the shifts from, first, the uterus to the ovaries as the main controller of female behavior and, second, to hormones as the messengers within the body that allowed scientists to achieve the more precise understanding that they had sought for centuries about the means of communication between female reproductive organs and the female brain.

In addition to these historical insights, the chapter's analysis has exposed another way in which topology functions as a way to explain the rhetorical movement of scientific concepts through time. In a manner that is similar to the way in which many early texts asserted that the womb was capable of traveling throughout the female body, expert ideas about femaleness have demonstrated a remarkable ability to travel through the centuries. This form of movement is captured quite effectively in contemporary rhetorical reconfigurations of remembering and forgetting, such as that offered by Stormer, but this understanding is also usefully enhanced by Serres's concept of percolation. The visualization afforded by the percolation metaphor enables us to see how the movement of remembering and forgetting specific beliefs about femaleness has been quite abrupt and unpredictable, and yet there is a liquid aspect to this movement that captures how even the oldest ideas can unexpectedly dribble into expert discourses at any given moment in history. In contrast to the previous two chapters, which focused on moments of stasis and then the dramatic movement that occurs at the moment when a stasis is unsettled, this chapter has examined a slower form of rhetorical movement that can occur as ideas travel through time. Of course, the particular set of ideas that I have explored in this chapter have been especially detrimental to women as they have made their way through time. On one hand, the female body was portrayed as very powerful, with the suggestion that women's minds and bodies had powers that far exceeded anything possessed by men's bodies. But, on the other hand, these powers were depicted as something to be feared; they made women's bodies harder to control, more dangerous, and more complicated than men's bodies were. And there was a persistent, underlying suggestion that women themselves were rational beings in the same way that men were because their bodies' mysterious powers caused their bodies to do things that they might not have even been conscious of.

This suggestion is expressed by Nebraska state senator Bill Kintner in a 2013 web interview:

> Biggest mystery? Women. No one understands them. They don't even understand themselves. Books and books and books have been written about it, and no one understands it.
>
> Men are very easy to understand. Very basic, very simple.[47]

Kintner's remarks, of course, provide further evidence of the remarkable ability of ideas to travel through time. But continuing this chapter's argument that such movement includes elements of both remembering and forgetting, Kintner's remarks also suggest an important way in which the ideas about sex difference that have been exposed in this chapter depart from the beliefs about hysteria that were examined in previous chapters. Although hysteria throughout history was believed to be largely a woman's disease, it was never exclusively seen as a female condition, and there have been many documented cases of males who suffered from hysteria throughout the ages. Furthermore, although there were times in history when hysteria was believed to be widespread among women of certain social classes and races, it has always been seen as a pathological condition that affected a subset of the population, not everyone. By contrast, the constructs of femaleness that are addressed in this chapter have been powerful and pervasive because they were intended to apply to all women; they were meant to suggest, as in Kintner's 2013 remarks, that women were fundamentally different from men and much more difficult to comprehend.

Although this chapter's main focus has been texts from a distinct period that stretches from the mid-nineteenth century to the early twentieth century, the theoretical framework that has been developed and employed in the chapter also helps us understand why some of these ideas keep percolating into current popular discourse. In response to public remarks such as the ones that have framed this chapter, there is typically a public outcry that leads the public figure to retract or apologize for his statement. For instance, after the public outcry that resulted from his remarks, Akin issued a campaign advertisement titled "Forgiveness." In this advertisement, Akin acknowledges the inaccuracy of his remarks about rape, and he asks his viewers for forgiveness. As he says, "the fact is, rape can lead to pregnancy" and "the mistake I made was in the words I said, not in the heart I hold."[48] Akin's campaign ad apparently came too late, as political opponents by that time had already used social media to ensure that his misogynistic remarks were made known across the world within hours of the moment that he uttered them. As a result, despite his public apology, Akin lost his 2012 congressional race by a wide margin to Democratic opponent Claire McCaskill.

Following Serres's notion that time progresses in a topological, rather than strictly linear, fashion, we can treat language such as Akin used in 2012 as a wormhole of sorts that grants us access to a curious mix of ideas that have developed over the centuries about what makes women different from men. This chapter's particular focus has been the prehormonal discourses of the nineteenth and early twentieth centuries, but as we have seen, the set of beliefs that crystallized into a distinct form during this period are beliefs that can be traced to ancient texts, and because of this long history, it is not unreasonable to expect that such ideas might continue to percolate back into expert and public rhetorics occasionally. Sometimes this resurfacing is quite explicit, as in Akin's outrageous remarks, but as I contend in this book, this resurfacing can sometimes be much more subtle, occuring even in scientific texts that are presumably objective and fact-based. Exposing the scientific basis for ideas such as those that were

articulated by Congressman Akin might not fully explain how such a prominent public figure could make these remarks with a straight face, but it gives us
reason to take the potential impact of these remarks more seriously. It also
reveals the line between the scientific truths of today and the nonscientific
beliefs of previous eras to be murkier than we might expect.

6

Illuminating Women | Metaphor and Movement After Centuries of "Groping in the Dark"

> In 1934, we all were excited over the immediate results: Nature had let us find one key to its secrets, and new horizons seemed to open up. But it was that first sight of the sparkling, glittering crystals in the retort that we will always remember—that moment of bliss for which the scientist would give a thousand days.
> —Adolf Butenandt and Ulrich Westphal, "Isolation of Progesterone—Forty Years Ago"

With this language from their 1974 article, "Isolation of Progesterone—Forty Years Ago," Adolf Butenandt and Ulrich Westphal reminisce about the moment in March 1934 when their team of scientists first isolated progesterone in the laboratory. This breakthrough was an important next step—following Allen and Doisy's 1923 isolation of estrogen—toward a complete understanding of the hormonal intricacies of the female menstrual cycle.

The authors claim in the article that they knew that they had been successful when "the biological assay in the infantile rabbit . . . showed that we had indeed isolated a compound with the highest progestational activity ever described; 0.75 mg. transformed the proliferative endometrium to the secretory phase." They go on to describe their initial excitement and the perceived significance of their discovery: "We had thus achieved our objectives: isolation of the corpus luteum hormone and elucidation of its structure. And we had opened the way to large-scale production, for application in human therapy."[1] In this chapter, I continue exploring this "elucidation" of the female reproductive system that came about through expanded knowledge of hormones in the early twentieth century. Continuing the previous chapter's emphasis on the ability of ideas to move through time, I expose additional ways in which hormonal explanations of the female body merged with older ways of thinking as these explanations became more deeply enmeshed in scientific understandings of the menstrual cycle. The key rhetorical concept in this chapter is metaphor. In the texts that constitute the focus of this chapter, I argue, hormones began to emerge more clearly as the metaphor that would transport ancient ideas about women's bodies into the twentieth century.

With the transition from the hysterical-woman metaphor to the hormonal-woman metaphor, we will see, physicians gained a much more robust vocabulary than was ever before available for describing the multiple ways in which women's bodies could have been pathological. Although hysteria was subject to many different definitions over the centuries, none of these ever facilitated a means of diagnosing women's pathologies that could have met increasing demands for the scientific support of medical claims and practices that were made by turn-of-century audiences.[2] By contrast, after the discovery of progesterone, in 1934, the hormonal-woman metaphor gradually gained ascendance over the hysterical woman that was her predecessor. As this new metaphor took hold, physicians gained access to the systematic vocabulary that was previously unavailable to them. As a result, hysteria was replaced by a variety of different diagnostic categories—including terms such as *premenstrual tension* and *premenstrual syndrome*—whose emergence will be explored in this chapter.

The word *metaphor* originates in part from the Greek word *phora*, which means "to transfer" or "carry over." Thus, an important concern of this chapter is to illuminate the precise forms of movement that metaphors can facilitate that occur in "the spaces situated between things that are already marked out—spaces of *interference*," in the words of Serres.[3] In particular, the chapter's analysis highlights four forms of rhetorical movement that were integral to this phase of transition from the hysterical woman to the hormonal woman as the dominant metaphor that was used to explain female problems: argumentative shortcuts, purposeful confusion, expansion of empirical evidence, and expansion of expert vocabulary for female pathology. Although these distinct forms of movement can be traced in key medical texts of the early twentieth century, they occurred simultaneously in an interconnected fashion, not necessarily as discrete forms of movement that were separate from each other.

Serres refers to Hermes, the god of communication, in characterizing the movement that is facilitated by metaphor in a manner that is useful as an entry to this chapter's analysis: "*Metaphor*, in fact, means 'transport.' That's Hermes's very method: he exports and imports; thus, he traverses. He invents and can be mistaken—because of analogies, which are dangerous and even forbidden—but we know no other route to invention."[4] Reflecting similar ideas, in rhetorical scholarship, much attention has been paid to metaphors that carry meanings across the borders that exist between popular and expert discourses,[5] between disciplinary divides,[6] or between historical eras.[7] In the transition from hysteria to hormones, we see a situation that is similar to the situation that Jensen has observed—namely, the historical evolution of metaphors that were used to characterize infertile women. The particular emphasis in these two situations is on metaphor's facilitation of movement through time. In the metaphorical shift from hysteria to hormones, this journey was not complete until much later in the twentieth century. However, by examining some of the earliest twentieth-century texts that invoked hormones, we can map its route and illuminate the female problems and symptoms that had been surrounded by so much darkness and mystery since the beginning of recorded history.

Hysteria and Hormones as Metaphors

To understand hysteria and hormones as metaphors requires an awareness of language as metaphorical in the broadest sense, in line with George Lakoff and Mark Johnson's assertion that "human *thought processes* are largely metaphorical."[8] Because metaphors can relate something new and surprising to something old and familiar, they often provide a powerful tool for teaching or explaining unfamiliar subjects to diverse audiences. For these same reasons, metaphorical language has been characterized as central to the production of new scientific knowledge. As Evelyn Fox Keller states, "making sense of what is not yet known is thus necessarily an ongoing and provisional activity, a groping in the dark; and for this, the imprecision and flexibility of figurative language is indispensable." Metaphors become powerful in science, according to Keller, when they "provid[e] explanatory satisfaction where it is not otherwise available."[9]

Medical language, in particular, has a metaphorical dimension that has been explored by several rhetorical theorists. Segal captures this idea in a way that is especially appropriate for understanding the hysterical woman and the hormonal woman as metaphors. Specifically, Segal suggests that diagnosis has become a metaphor, or "an idea in which we have invested a series of meanings transferred from another medical idea, the idea of health itself."[10] In other words, a diagnosis is an explanation and the point of connection between physicians' and patients' understandings of the health conditions that leads patients to seek medical attention, and it facilitates transfer of information from the

expert to the nonexpert in a way that will hopefully enable each of these individuals to take whatever steps are necessary to achieve health and wellness for
the patient. We can extrapolate from Segal's understanding of diagnosis as
metaphor to consider how hysteria and hormones developed, each in their
respective eras, as metaphors that explain female problems to patients themselves, to physicians, and to the larger communities in which we live. As metaphors, we will see, these terms have enabled and constrained the production of
knowledge about the various health conditions that they are invoked to explain,
and they have been implicated with distinct forms of medical treatments and
approaches for dealing with these conditions. Tracing these metaphors as they
evolved over several centuries reveals a number of different ways in which the
hysterical- and hormonal-woman metaphors each facilitated particular kinds of
rhetorical movement, sometimes acting individually and sometimes acting
together to "coalesce and form entirely unique mixed metaphors composed of
metaphors-of-old."[11]

Menstruation as Pathology: Some Early Ideas

We have seen in previous chapters the extent to which the hysterical-woman
metaphor "provid[ed] explanatory satisfaction"[12] to physicians, patients, and
public audiences for the problems that were specific to women and had long
been surrounded by confusion, mystery, and darkness. In the context of menstruation, as in other aspects of women's health, we see an example to illustrate
Segal's assertion that diagnosis is metaphorical. Specifically, the hysterical
woman served as a metaphor that caused medical experts, and women themselves, to understand and account for female problems in a particular way. As we
have seen, this metaphor dominated the diagnosis of a wide array of female
problems for many centuries, although the particulars underwent many changes
throughout the eras. Thus, in some instances, a failure to menstruate was depicted
as a cause of hysteria. For instance, the author of *Aristotle's Masterpiece* devotes
much attention to the adverse effects of "retention of the courses," which he said
could include bouts of hysteria.[13] But sometimes menstruation itself was seen as
causing hysteria or exacerbating its symptoms. For instance, Veith's historical
work on hysteria reveals that as early as the eighteenth century, physicians were
writing more precisely and systematically about the cyclical nature of hysterical
symptoms, including migraines as well as more severe hysterical attacks.[14] These

physicians depict a confusing relationship, and how they conceived cause and effect is not entirely clear. As Veith says in regard to one of these physicians, "menstruation is mentioned, for the symptoms become aggravated at the periods, which were generally believed to augment emotional tensions even in normal women." This physician also "insisted that all local diseases of the uterus, ovaries, and vagina were likely to be followed by hysteria, which then may gradually progress into insanity."[15]

Thus, for these eighteenth-century physicians, the cyclical nature of hysteria symptoms remained a mystery that could not be solved while the hysterical-woman metaphor dominated physicians' diagnostic practices. As hormonal explanations for the female menstrual cycle emerged in the early twentieth century, however, hysteria gradually slipped away and was replaced by descriptors such as *premenstrual tension*. With emergent terms such as these, we start to see a shift from the hysterical woman to the hormonal woman as the dominant metaphor in physicians' diagnoses of female problems. Even as the metaphor slowly began to shift, many of the symptoms stayed the same; these included physical problems such as migraines and indigestion as well as nervous symptoms such as anxiety, shortness of breath, and seizures. But we see evidence of the shift in metaphor when we look closely at the different explanations that medical experts used to account for these symptoms. A 1927 article by Chauncy Allen offers a glimpse at the early phases of this shift in terminology that started to occur with the gradual transition from the hysterical-woman metaphor to the hormonal-woman metaphor. In a few places throughout Allen's article, we see the mixing of metaphors that is characteristic of scientific language, especially when beliefs shift.[16] Because scientific inquiry involves exploring an area or subject that is fraught with uncertainty—what Keller calls "a groping in the dark"—one element to scientific metaphors is that they allow for remnants of the old ideas to persist in the new ones; as Keller says, metaphors are one of the entities that enable scientific narratives to "embed earlier meanings in new constructions."[17]

Reflecting these insights about mixed metaphors, Allen's article contains language about both hysteria and hormones, and throughout the article, he mixes these two metaphors in unpredictable ways. Thus, for instance, early in the article, Allen cites previous research that demonstrated that women were "more than usually suggestible" at the time of menstruation. The language he uses to explain this behavior recalls the hysterical-woman metaphor in its emphasis on the interaction between the uterus and other organs: "A great deal of the pseudo-menstrual suffering reported was simply coincident functional disturbances in other organs." However, as he proceeds with this discussion of previous research, Allen mentions "endocrine secretions, especially those from the gonads," as one possible factor in these symptoms that women experience. He does not provide great detail about this cause-and-effect relationship, and he ultimately reverts to hysteria as an explanation, quoting a psychoanalyst who had recently claimed "that *every* case of spasmodic dysmenorrhea is an anxiety hysteria, and is curable by suggestion and an understanding of the individual case." Allen then echoes what had been a long-standing belief about the psycho-somatic nature of hysteria symptoms, concluding that the pain and discomfort of menstruation had only existed in women's minds and could have been con-trolled if women had been taught to change how they thought about the condi-tion: "Menstruation—on the great average—is far less a handicap to woman than her warped idea of its effect is, and that she performs mental and motor acts almost, if not equally, as well during her menses as when she is freed from this seeming disadvantage."[18] In these first few pages of the article, Allen shifts back and forth so frequently between hormones and hysteria that it is some-times hard to tell which metaphor he is using to express a particular idea. In this language, we start to see how hysteria was not replaced by hormones at a single time and place; instead, it is more accurate to say that those experts who contin-ued to cling to the metaphor of hysteria molded and shaped the metaphor to correspond with other ideas that took hold at this time—most notably, the idea of the hormonal woman.

As we progress through the early twentieth century, this mixing of meta-phors continues, but physicians' explanations started to depend more on hor-mones as hysteria faded into the background. For instance, in chapter IX of his 1929 book *The Female Sex Hormone*, Frank expresses ideas about menstruation that are similar to Allen's ideas.[19] Even though this book was published only two years after Allen's 1927 article, however, Frank relies exclusively on hormones to explain the pathological effects of menstruation. Basically, he describes all the phases of a "normal" woman's reproductive cycle, from the menstrual cycle to gestation, childbirth, and lactation. His descriptions of these phenomena cor-respond with how we understand these phases today, but he uses the generic term *female sex hormone* to designate the substance that causes these changes in the cycle to occur. This is likely one of the first medical texts, or at least one of the most prominent, to offer an explanation of women's problems that was based on hormones as the primary cause of these problems, although Frank's vocabulary for discussing these hormones is limited by the fact that progesterone had not yet been identified when he was writing.

A couple of years later, in a 1931 article, Frank first used the term *premenstrual tension* to describe these symptoms.[20] Frank's article elaborates on the general ideas introduced in his book and provides a more specific explanation of the role of hormones in premenstrual tension. The article is based on a series of case reports about patients who suffered "grave systemic disorders" during the premenstrual phase—that is, symptoms that went beyond the "fatigability, irritability, lack of concentration, and attacks of pain" that he says were typical during the premenstrual phase. Although he continues to describe these symptoms as having resulted from hormones, hysteria was not entirely absent from Frank's explanations. In fact, the patients in his case reports suffered serious events that sound quite similar to the symptoms that were previously affiliated with hysteria. For instance, the first patient reportedly suffered "frequent convulsive attacks" during the period "within ten days preceding menstruation." Neurological examination of this patient confirmed a diagnosis of "idiopathic epilepsy." The physician reportedly used a series of "roentgen treatment to the ovaries" to inhibit hormone production, and this treatment eventually caused the patient to go into "a period of remission." The second patient suffered from "severe bronchial asthma," which became more severe during the premenstrual period. This patient was also treated successfully with a "sterilizing dose of roentgen ray . . . directed against the ovaries." In summarizing these two cases, Frank uses language that is surprisingly similar to that which was previously used to account for the influence of the ovaries or uterus on other organs in hysteria patients: "These two cases illustrate well the close connection between the ovarian function and systemic manifestations due to other organic systems."[21] We will see, however, that in the remainder of his article and subsequent articles that build on Frank's research, the shift from the hysterical to the hormonal woman became more pronounced. As this happened, the metaphorical shift enabled the four additional forms of rhetorical movement that are the focus of this chapter: argumentative shortcuts, purposeful confusion, expansion of empirical evidence, and expansion of expert vocabulary for female pathology. Close examination of these four forms of movement as they occurred in key medical texts of the early twentieth century deepens our understanding of the rhetorical implications of the early twentieth-century shift from hysteria to hormones as the metaphor dominating the medical diagnosis of female problems.

Argumentative Shortcuts

The shift to the hormonal-woman metaphor becomes more pronounced in the remainder of Frank's 1931 article. By employing the hormonal-woman metaphor, he is able to provide a much more precise explanation of the physiological events that were involved in these phenomena than was evident in his earlier, ovary-based explanation. He says, for instance, "these estrual phenomena are dependent on the action of the anterior lobe of the pituitary which liberates a hormone that circulates in the blood." As support for this claim, Frank refers to scientific research that was relatively new at this time: "The hormones that produce these phenomena have been and can be recovered from the blood and urine, and in the case of the female sex hormone, have been studied quantitatively under normal and abnormal conditions."[22] Similar language appears in his 1929 book, in which, for instance, he says, "the demonstration by Philip Smith and by Zondek and Aschheim of the dominant influence of the anterior pituitary lobe upon the genital system, through its influence on the ovaries, has actually afforded proof of the long suspected and constantly referred to interaction upon each other of two glands of internal secretion."[23] In passages like these, we see how the hormonal-woman metaphor enabled a more precise account of the brain-to-sex-organ relationship than that which had been postulated in vague, nonscientific terms that were supported by the hysterical-woman metaphor many centuries prior. The specific chemical reactions that hormones enabled inside the female body received much research attention at the time when Frank was writing, and this trend continued through the next several decades.

Another component of Frank's assertion, however, that is simply assumed to be true yet not supported with any scientific evidence is present in Frank's suggestion that the physiologic events that hormones produced inside the body led to specific female behaviors and symptoms. For instance, in his article Frank describes, at great length, the serious effects of these hormones on the subset of women who concern him in this article. Frank's language in this part of the argument is far less precise, and it is based purely on his own interpretation of the patients' reports of their symptoms:

> The group of women to whom I refer especially complain of a feeling of indescribable tension from ten to seven days preceding menstruation which, in most instances, continues until the time that the menstrual flow

occurs. These patients complain of unrest, irritability, like "jumping out of their skin" and a desire to find relief by foolish and ill considered actions. Their personal suffering is intense and manifests itself in many reckless and sometimes reprehensible actions. Not only do they realize their own suffering but they feel conscience-stricken toward their husbands and families, knowing well that they are unbearable in their attitude and reactions. Within an hour or two after the onset of the menstrual flow complete relief from both physical and mental tension occurs.[24]

This passage illustrates how mixed metaphors conflated hysteria symptoms with new symptoms that came to be seen as the result of hormonal fluctuation, and it also suggests one of the key forms of rhetorical movement that were enabled by this mixing of metaphor—namely, a form of movement that would enable experts in subsequent generations to make increasingly sophisticated arguments about hormones as the basis for women's pathologies without requiring a great deal of scientific evidence to support their claims. There was a disparity, in this regard, between the amount of scientific evidence that was used to support claims about the physiological effects of hormones and the amount of scientific evidence that was used to support claims about the psychological effects of hormones. Because of its connections with the hysterical-woman metaphor, the newly emerging hormonal-woman metaphor enabled these early twentieth-century texts to make many claims that hormones caused dangerous, problematic psychological behaviors and symptoms for women in a way that did not require any evidence to support these claims; physicians accepted these psychological claims at face value, and they expected their audiences to do the same. Because for so many centuries it had been assumed that something inside women caused them to behave less rationally, more erratically, than men did, this part of the developing argument about hormones' effects on women did not need to be investigated but, rather, could be taken for granted. It mattered little whether the something inside women that caused these behaviors was understood to be a wandering womb, a problem in the ovaries, or hormones.

Purposeful Confusion

Next, Frank discusses normal and abnormal hormone levels in individual women who have been his patients, and we see that the hysteria-hormone mixed metaphor enabled another form of rhetorical movement: an oscillation between the assertion that some women were pathological and the assertion that all women were pathological.

For example, Frank mentions a thirty-two-year-old woman who was a patient. He begins by describing symptoms that could have previously been affiliated with hysteria, even making an explicit reference to what Charcot used to call "hystero-epileptic" symptoms. For this woman, "occasionally, eight weeks elapsed before menstruation occurred, under which circumstances hystero-epileptic manifestations occurred." The woman's symptoms were apparently so severe that "the patient was ready to commit suicide, and from her own description it was evident that a family rupture was imminent, although she said that her husband was both considerate and kind." After this initial use of terminology affiliated with hysteria, though, Frank shifts to the terminology and diagnostic techniques that only became available after the emergence of hormones. These included testing hormone levels in the woman's blood, and not surprisingly, he found "twice the amount of female sex hormone that is normally found premenstrually." As with the other patients whose cases are reported earlier in this article, Frank reports that "it was decided to tone down the ovarian activity by roentgen treatment directed against the ovaries." For this patient, further blood tests after the ovarian treatment revealed that "less female sex hormone than normal was now present," and "the nervous disturbances of this patient have been relieved for three years." The physician's explanation is that "an excess accumulation of hormone caused the symptom complex complained of, and could be temporarily relieved by venisection and permanently improved by reduction in the amount of female sex hormone in the circulation."[25]

After this detailed narrative of a single patient's case, Frank includes a table that summarizes the symptoms, treatments, and outcomes of several patients' cases. In the "Complaints" column of this table, he includes such reports as "suicidal desire," "unbearable, shrew," "husband to be pitied," "almost crazy," "psychoneurotic," "incapacitated mentally," "impossible to live with."[26] It is difficult to discern exactly how he obtained these patient reports, but many of the descriptions are in quotation marks, suggesting perhaps that the language came from the woman's husband or another family member.

Frank then explains that the symptoms that he discusses in this article can be attributed to "continued circulation of an excessive amount of female sex hormone in the blood," and the resulting "periodic attacks are incapacitating and lead occasionally to extreme unhappiness and family discord."[27] This is one of the first articles to establish such a direct connection between hormone levels as measured in the bloodstream and the kinds of problematic behaviors and symptoms that had previously been assigned to hysteria. These symptoms "can be directly ascribed to the excessive hormonal stimulus." This language is important because Frank asserts here that excessive female hormones—not just regular levels of the hormones—led to problems. Finally, Frank confirms the hormonal explanation of these problems in his statement that the most effective treatment for severe symptoms was to reduce hormonal activity in the female body by treating the ovaries with radiation: "At present, in the severest cases of this nature temporary or permanent amenorrhea, brought about by roentgen treatment, appears to be the proper procedure."[28]

The last few pages of Frank's article comprise a section titled "Abstract of Discussion." This section reports the follow-up discussion among audience members that took place after Frank presented this paper at a conference. In this section, we see the purposeful confusion that is enabled by a mixing of hysterical and hormonal metaphors. Specifically, in some physicians' comments, it is suggested that it was not just abnormal hormone levels that made some women pathological, but rather, it was presence of the female sex hormone at any level that made all women pathological. For example, the first physician quoted in the article, Dr. Josephine H. Kenyon, reports that these premenstrual symptoms posed a problem for all of her female patients, not just ill ones: "Premenstrual tension is an outstanding complaint of the presumably well women who present themselves periodically for examinations."[29] Then another physician, Dr. Edith R. Spaulding, responds by establishing a relationship between these premenstrual symptoms that presumably appeared in normal women and the symptoms that had long been affiliated with the pathological condition of hysteria: "One is reminded at once of the hysterical patient who is unable emotionally to cope with her whole sex life, and who shows uncontrollable tension during the premenstrual period. In many instances in that type of case, the feeding of ovarian extract produces an exacerbation of the emotional symptoms. This reaction, while not constant in all cases, seems to occur in the cases in which overstrong sexual emotions have found outlet in hysterical symptoms."[30] It is hard to understand exactly what relationship this physician believed existed between hysteria and premenstrual tension, but hysteria had for so long been understood in so many conflicting ways, and had been confused with so many different conditions anyway, that it is not at all surprising that there would be some confusion between the older understandings that attributed all of women's problems to the pathological condition of hysteria and the newer understandings that explained these same symptoms as the result of hormones, even in perfectly normal women.

Spaulding then responds directly to Dr. Frank's recommended treatment of using roentgen rays to stop the ovaries from producing hormones, and her language offers another important example of confusing normal women with pathological women that occurred with the mixing of hysterical- and hormonal-woman metaphors. Specifically, in this physician's comment, we see how female pathologies have been related to, and sometimes have been confused with, the social or interpersonal problems that women also experience. As Spaulding says, "the treatment outlined by Frank suggests an interesting possibility in such cases. On the one hand, the patients can be helped psychologically to utilize their increased sexuality through gaining a better understanding of themselves and finding a more adult outlet for their emotions; on the other hand, in many cases, especially in unmarried women with deep-seated emotional difficulties, the lessening of the sexual drive offers a helpful therapeutic procedure."[31] It is unclear what this physician might mean by helping patients "utilize their increased sexuality," but the key point is that women's sexuality, as it had been for many centuries prior, is treated here as something inherently problematic. In this case, the problem seems to have been women who possessed too much sexual drive, whereas in many other cases, it seems to have been the opposite problem—that is, too little sexual drive. But the important observation is that women's sexual drive created the health problems that at this time were coming to be affiliated with premenstrual tension; this is an important point of connection between the newer hormonal explanations and the older explanations that were based in hysteria, and it is another way in which the hormonal-woman metaphor left open the possibility that all women, not just a subset of them, could be construed as pathological.

In these apparent contradictions in Frank's article, we can see how the mixing of metaphor allows for various forms of confusion between the belief that female hormones made all women pathological and the belief that it was only abnormal levels of female hormones—which could be measured in the bloodstream—that made some women pathological. It is not entirely clear which of these views Frank espouses in the article, but it is possible that his language allows for a purposeful confusion between these two meanings. This potential for confusion or obfuscation has been documented in previous studies of mixed metaphors, particularly in the context of genetics, where it has been shown that mixed metaphors, both in scientific and popular discourse, can sometimes obfuscate and sometimes clarify the way that we understand how genes impact human health.[32] As we will see, in the case of the hormonal-woman metaphor, physicians who came after Frank benefitted from this ability to oscillate between the assumption that all women are fundamentally abnormal (because they have female hormones) and the assumption that only some women are abnormal (because their hormone levels are too high or too low).

Expansion of Empirical Evidence

As the hormonal woman gained prominence in physicians' explanations of female pathology, physicians strengthened their efforts to provide empirical evidence for their assertions about the effects of hormones on women's behaviors. R. E. Whitehead's widely cited 1934 study of women pilots is an example of such efforts. This study was one of the first to offer empirical evidence for the long-standing assumptions and observations that women's mental and physical performance were diminished during menstruation. Whitehead's article presents a series of written reports about women pilots who crashed their planes while they happened to be menstruating. In the first story that he quotes from, the woman pilot says, "I was feeling unwell (menstruating) that day and should not have gone up, but wanted to practice a little as I was supposed to complete the test for private pilot's license the following day at _____ Airport. I had made two landings and had cut the motor at about 1,000 feet to make a third landing. I then *fainted* and from what the spectators say I must have stalled it and it spun into the ground. It was no fault of the plane."[33] This paragraph has several key rhetorical features, but perhaps the most important is the extent to which the woman blames herself—that is, a deficiency or weakness in her own body—and asserts explicitly that the crash had nothing to do with a possible malfunction of the plane. Furthermore, it is interesting to note that she highlights fainting as a symptom. Fainting, of course, hearkens back to Victorian images of women passing out and the use of smelling salts to revive them, which hearkens back to hysteria and attracting the wandering womb back to its correct position.

The woman pilot in the next story died in her crash, so she was not available to narrate the series of events. However, the author of the article confirms that the crash was caused by pilot error: "Miss _____ hit the ground about 20 feet from the plane and when found was still holding the handle of the rip cord in her *left hand*. (She should have used her right hand)." And, of course, "further investigation by Coroner _____ reported that Miss _____ was in her menstrual period." Additionally, "her friends, who were with her the evening preceding the accident, stated that Miss _____ was in a highly nervous condition."[34] Once again, language such as this offers extremely useful insights into the shift that occurred in the early twentieth century as vague notions of hysteria came to be replaced with the more scientific notion of hormones.

Whitehead reports another accident in which the woman pilot died: "The Coroner reported that Mrs. _____ had taken several aspirin tablets prior to participating in the race, for a headache resulting from her regular menstrual period." Of course, it is not clear how a coroner would be able to determine such a precise reason why a woman had taken these aspirin, but that was his conclusion, nonetheless. Whitehead admits that there were two possible causes for this crash: "It is possible that the pilot fainted or that the plane became uncontrollable due to the rib stitching cutting through the rib stitching tape and fabric on the entire upper side of the right wing, allowing it to balloon." He does not state which of these was the actual cause. However, in the next section of the article, he summarizes the correlation that has been observed between menstruation and women pilots' accidents: "This [correlation] may be a coincidence, but since we have had a number of cases we are beginning to think that it is not." Oddly, he then says, "some localities in the United States have been practically depleted of women pilots by accidents." He then draws a confusing comparison between men pilots and women pilots: "What a man is liable to do from one moment to another is apparently beyond the control of outside forces and this makes it often difficult to prevent accidents," and this same observation "may apply to women flyers as far as their menstruation is concerned." He admits that women had the same kind of individual variation that he had earlier assigned to men: "Some women can probably fly with perfect safety during their menstrual period while others cannot." This was because "the majority" of women would probably have been able to know the difference and would have made the responsible choice not to fly a plane during menstruation if they were not capable of doing so safely. However, he says, not all women were capable of making this distinction, and "if they are not able to evaluate their frailties while on the ground, how are they going to evaluate them in the air?" Whitehead's conclusion, then, is that women need to be educated about the possible negative effects of menstruation on their abilities as pilots: "Although many women hesitate to speak frankly about the effect menstruation has upon their emotional and mental make-up, still the importance of this subject should be brought to their attention."[35] This relates to the recurring theme that the knowledge of hormones' effects on behavior and mental health originated with men. Women were basically passive objects to be observed and examined; they were not active participants in this observation and the creation of knowledge that derived from it.

Another 1934 article, "The Female Sex Rhythm," is also important in demonstrating how the hormonal-woman metaphor came to be enmeshed with longstanding beliefs about the menstrual cycle as mentally and physically debilitating. This review article's author, Georgene H. Seward, cites a diverse body of literature that includes studies about women as well as other mammals. The author still speaks quite vaguely of the "female sex hormone," and she perpetuates the trend of not providing any scientific support for claims about hormones' effects on women's behaviors. Seward begins by citing literature that shows how the vital signs—including pulse, blood pressure, and body temperature—varied in response to hormones during the menstrual cycle. Next, she shifts to the effects of the menstrual cycle on women's nervous systems. The studies that she cites are quite specific, examining, for instance, how women's field of vision changed throughout the menstrual cycle. This is where we see the dramatic changes that occurred in scientific understandings of women's bodies after the discovery of hormones. However, there is language in the article that is strangely similar to the language that had been used by experts in earlier generations to describe the causes of hysteria. Seward even reports a case of "visual disturbance occurring during menstruation and pregnancy," which was attributed to "hormonal toxic influences." She then discusses "cutaneous sensitivity," and she suggests that the female sex hormone "increases the sensitivity of the cerebrospinal neurones to incoming impulses." Next, in studies of reflexes, she cites a study that found "a period of hyperexcitability in connection with menstruation but bearing no *constant* relationship to it."[36]

Next, Seward discusses "spontaneous" activity, with a particular focus on the white rat, which she says is a good subject "because of the spontaneous rhythms of activity it displays." She discusses several rat studies, including studies of "oestrus rhythm" in the rat and its effects on the female rat's sex drive. In the next section, on "Motor Coordination," she shifts back to discussing women. On this topic, she begins by discussing research that had apparently been conducted to test a woman's balance. She describes a "general tension and restlessness" that was observed during the premenstrual period. She says the author of this study "regards the restlessness as a distraction of the attention from the necessary kinesthetic cues." Extrapolating from this author's conclusions, Seward adds her own hormonal interpretation of this phenomenon, suggesting that "both the restlessness and the distraction are effects of underlying endocrine disturbances." She also cites a similar study conducted on men, which found no variation or fluctuation based on hormones or cycles. Next, she discusses a study that required women to walk on a tightwire as their performance was correlated with the phase of their menstrual cycles: "There was a marked impairment of efficiency at the menstrual phase of the cycle, followed by a relatively rapid improvement with a peak at about the 11th day after the end of the period." She then moves back to a study of "maze learning in the white rat."[37] The study seems to be inconclusive, failing to show any effects of the estrous cycle on the rat's performance in the maze.

Seward moves on to "psychomotor efficiency," shifting back to human women instead of rats. She discusses a series of tests of "simple and choice reaction time [and] continuous free association" in addition to a "Vaschida test for attention, Kraepelin arithmetic test, digit span, and memory [test] for discrete words." This study apparently showed several measures that were negatively affected by menstruation, including "the stream of free associations" and "attention and arithmetical efficiency performances."[38]

The next passage in Seward's article is important because it exposes the underlying assumptions that motivated much of this research and provides a clear illustration of the manner in which the hormonal metaphor supported explanations of women as pathological: "On the basis of these results, the author concluded that the menstrual function does not constitute a work handicap in the normal woman. Although this experimenter introduced the necessary external controls, she did not allow for the internal variable of compensatory effort that may have made it possible for her subjects to overcome any disadvantage of the menstrual condition for the brief duration of these tests."[39] This passage suggests the extent to which Seward expects that there must have been some negative effect induced by menstruation on the woman's work performance, and if that effect did not show up in the study results, then that must have been because the woman was working secretly to overcome it. Rather than viewing that ability as a strength that indicates that women could succeed in any circumstances, however, it is viewed in this study as a "compensatory effort"—in other words, an additional effort that was necessary just to bring her performance up to a (presumably male) norm that defined what was a minimum expectation.

Seward moves on to discuss research about "Affective Reactions" to the menstrual cycle. She first cites a researcher who postulated that the main difference between the menstrual and nonmenstrual phases of a woman's cycle was different levels of "mental satiability." This research supports a speculation that "the most significant cyclic variations may lie in the affective rather than the intellectual spheres." She suggests that this was an important area for future research, noting that at this time there was "practically no experimental work on the effect of the menstrual cycle on feeling and emotion."[40]

Whitehead's and Seward's articles, both published in 1934, seemed to be important texts in establishing the future trajectory of research about behaviors and symptoms that would come to be more fully accepted in the medical community as hormonal effects that were related to phases in the menstrual cycle. Because Seward's article is a comprehensive review, it is not surprising that subsequent researchers would cite it as an authoritative source. It is interesting, however, that Whitehead's three-page article about women pilots has received almost as many citations as Seward's forty-page review article. Whitehead's article is entirely based on a series of case reports, sometimes using the woman pilots' own language but sometimes using a (usually male) expert's language to narrate the events that led to the accident, such as in situations in which the pilot died in the crash. Whitehead does not cite a single scientific source to support any of his claims, and in fact, his article is in a section of the *Journal of Aviation Medicine* that is titled "Notes from the Department of Commerce." Despite these limitations, this three-page article was cited, alongside Seward's comprehensive review, for many decades as an authoritative source for claims about the effects of menstruation on women's physical and cognitive abilities. Both articles were cited in notable venues such as the *British Medical Journal* and *Psychological Bulletin*.

Expanded Expert Vocabulary for Female Pathology

Therese Benedek and Boris Rubinstein's two-part study of "The Correlations between Ovarian Activity and Psychodynamic Processes," published in 1939, is one important example of a study that made further advances, beyond Whitehead's and Seward's, in providing empirical observations about the hormonal effects of menstruation on women. We also see, in this study, how the more precise vocabulary for female pathology that emerged with the hormonal-woman metaphor allowed experts to move beyond earlier concepts such as instinct and perception that had often been used to characterize female cognition when the hysterical woman was the dominant metaphor. Benedek and

Rubinstein's articles are also significant because they were among the first to integrate the identification of progesterone into their understanding of the menstrual cycle, thus enabling a more complete understanding of this cycle than was possible before. This also enabled these researchers to take empirical observations like those reported by Whitehead and Seward and report them as facts that could be verified with both laboratory and clinical research.

The opening paragraphs of part 1 of this study suggest how Benedek and Rubinstein positioned their study in relation to the long history of beliefs about female problems: "The existence of relations between gonadal function and emotional states had been inferred before the dawn of history and can be traced in the folklore of nearly all people. . . . The ebb and flow of emotion in the adult woman has also been associated in a vague way with the cycle of sex function." The authors cite Seward's 1934 article on the female sex rhythm as a reference for this claim, and they proceed to describe psychological conditions such as "premenstrual nervousness" that, they report, had been documented for years. But they claim that their study is the first to offer verifiable clinical evidence to explain the cause of these conditions: "Proof of such correlations has, however, been strikingly absent due to ignorance concerning the precise details of the cycle in women on both the physiological and psychological sides."[41] The authors then cite a few studies that examined the physiological side through vaginal smears and basal body temperature recordings. They note that, in their own study, "the psychoanalytic method" enables a consideration of the psychological side.

In the next few paragraphs, Benedek and Rubinstein describe the female reproductive cycle, starting with ovulation and moving through menstruation. Benefitting from the 1934 identification of progesterone, the contemporary scientific understanding of the entire menstrual cycle was more complete than it had ever been before. Next, the authors announce more specifically their study's contribution to previous research: "The existence of a hormonal cycle which is reflected in the vaginal smears and basal body temperature has been established. It was, therefore, interesting to see whether the psychological material could be correlated with the hormonal purpose." The authors summarize results of the two different test methods in a table that includes two columns: "Prediction on the basis of psychoanalytic material" and "physiological findings." The left-hand column, which summarizes psychoanalytic material, uses language such as "heterosexual tension," "preovulative tension," and "increased premenstrual tension." In this left-hand column, the researchers report what they predicted about each participant's psychoanalytic state, based on verbal material such as dream transcripts that the participants wrote. In the right-hand column, the authors report physiological data that indicates, through measures such as vaginal smears, which phase of the cycle each participant was in.[42]

After the results of the physiologic and psychoanalytic evidence are reported in tabular form, the authors analyze and discuss the results. They assume a distinction between normal and abnormal responses to hormones and equate normal behaviors and mental states with normal sexual activity in a manner that clearly echoes ancient ideas about the womb as a wandering animal that sought to be satiated with sexual intercourse. The following language typifies this pattern of reasoning: "With normal sexual adjustment the increasingly strong heterosexual desire finds normal gratification. Without sexual gratification, the heterosexual tension can be dammed up so that increasing hormone production causes an increased tension." Another passage is even more interesting because it addresses the "neurotic" woman, but it does so in a way that suggests that being a neurotic woman meant being a woman who was not content with her femaleness. This passage also leaves room for a possible interpretation that we have seen in other texts—namely, that at least during certain times of the month, to be a woman was to be neurotic: "In neurotic persons we observe that this increasing oestrone production activates the psychological conflicts and thus the neurotic symptoms are intensified. The great psychic tension is suddenly relieved . . . when ovulation occurs. . . . She is self-satisfied, wants to be loved and to be taken care of. She is content to be a woman."[43] Later in the article, the authors weave together the various pieces of evidence to support what they call "the instinct theory of psychoanalysis." They claim that their experiments "offer the first laboratory results to support the theoretical concepts of psychoanalysis," and then they proceed to introduce "the terminology of the instinct theory."[44] This terminology consists of several different kinds of libido (oral, genital, active-passive, etc.). The instinct theory, as related to hormones, is summarized as follows: "The instinctual tendency changes its direction after ovulation took place." More specifically, "while it was active, and directed toward the sexual object during the follicle-ripening phase, it becomes passive and directed toward one's self after ovulation."[45] Further still in the article, it again becomes clear that this connection between hormones and behavior was specifically a female phenomenon. As the authors explain it, the problem of being a woman was tied in a particular way to the flux of hormones throughout her reproductive cycle: "Progesterone is the hormone chiefly concerned with preparation of the uterus for nidation and with maintaining pregnancy. The physiological preparation of the uterus for nidation implies a task for the psychic apparatus to be dealt with in every cycle, namely to *solve the problem of being a woman"* (emphasis added).[46] This is the second time in the article that the authors refer to the condition of being a woman as something that was inherently pathological, implying that it was a problem that individual women had to learn to overcome if they were to be mentally healthy. This is, of course, not a new idea; it can be traced to the ancient texts that found various ways to depict women as defective or inferior versions of men. However, what is new about these ideas as articulated in this study is that the researchers provide the clinical evidence that is needed to understand a vague concept like that of instinct as a scientifically verified phenomenon. It becomes quite clear in this article how hormones provided the missing puzzle piece that finally allowed claims about women as inherently pathological to gain scientific credibility. And it is important to note that these authors do not use the term *hysteria*; rather, they seem to replace *hysteria* with *neurosis*. The hormone-behavior connection is stated most clearly in the conclusion section of Benedek and Rubinstein's article: "The investigation suggests that in the adult woman, it was possible to relate instinctual drives to specific hormone functions of the ovaries." Their concluding statement is even more revealing of the manner in which this scientific evidence of hormonal fluctuation is woven into previous theories that explained women's behaviors through vague notions like instinct: "This method affords an approach to the study of the biological foundation of instincts."[47]

In part II of this study, the opening paragraph further suggests how the idea of instincts transformed when hormones became part of the language that scientists used to talk about women's brain functions. The authors' language is much more sophisticated than the language of earlier texts had been, but Benedek and Rubinstein still use the two different terms, *instincts* and *hormones*, together: "In the first part of this communication we presented evidence indicating that human instinctual drives are controlled by gonadal hormone production." This article's conclusions are based on the evaluation and comparison of "endocrine and psychoanalytic records." Specifically, the authors collected vaginal smears and basal body temperature data to determine the time of ovulation, and they then sought to determine "correlations in the psychological and physiological processes on each day of the cycle."[48]

This article typifies the more scientific approach to connecting female hormones and brains that we are now accustomed to, and in fact, it articulates many ideas that, we will see in the next chapter, persist in twenty-first century scientific texts. Benedek and Rubenstein are no longer basing their claims about sex difference on dissected brains, and they are not even concerned with the brain's anatomy. Rather, they construe mental capacity as part of the central nervous system and as fundamentally connected to and manifested in other behaviors, such as sexual desire, that research subjects addressed in dream transcripts. Thus, for example, the following passage describes something that sounds similar to what we now call *premenstrual syndrome*. This passage sounds more like today's explanations because it refers to central nervous system activity and depicts the human body as a "complex system,"[49] with all its parts interacting unpredictably, rather than focusing on specific interactions between discrete organs: "The hypothetical mechanism of the precipitation or suppression is probably through hypothalamic autonomic centers. This suggests that menstruation involves neural as well as endocrine mechanisms." The authors go on to describe "the onset of flow" as a "hormone withdrawal symptom." Later in the article, the authors switch to a more explicit focus on the hormone-brain connection, but they proceed gradually. They make one explicit reference to instinct: "We correlated the specific hormonal functions of the ovary with specific instinctual drives." When they talk about instinctual drives, they seem to be talking about different kinds of sex drive—an active heterosexual drive in the first half (estrogen-dominated) of the cycle and a more passive receptive drive in the second half (progesterone-dominated) of the cycle. The methods that they used were quite complicated, and they essentially relied on pathological women, or what they call "neurotic" women, as their test cases. The authors explain that "it is easiest to make quantitative comparisons in neurotic women whose inhibited and dammed-up instinctual drives call for strong defense reactions." They mention such traits as aggression and anxiety as manifestations of these defense reactions. Although they briefly discuss how hormone fluctuation influenced the behavior of so-called "normal" women, they continually emphasize that this correlation was harder to trace in normal women: "The ovulation phase in such [normal] women is psychologically not so obvious as in our case material of neurotic women."[50] Interestingly, the authors mention "sociological factors," but they say that these are outside the scope of this paper; instead, they focus exclusively on the effects of hormones on women's behaviors as manifested in psychoanalytic records. Using a method similar to that employed in part I of their study, the authors report that their study "confirms the probability that in the adult woman, instinctual drives are related to specific hormone functions of the ovaries." They also conclude that "there is a semi-quantitative correlation between hormone production and psychic tension."[51] The authors cite a number of sources, the earliest published in 1925. These sources report similar research, and they reflect similar assumptions about the relationship among hormones, the brain, and sex drive. There is not much talk about the brain as an organ, but there is more general discussion of "psychic" and "neurotic" responses.

Emergence of Premenstrual Syndrome

By the mid-twentieth century, Benedek and Rubinstein's 1939 ideas were accepted as scientific facts, and they had led researchers to develop an even more precise vocabulary for female pathologies. A 1954 article by Raymond Greene illustrates this development. The transition from hysteria to hormones was even more complete at this time because Greene had access to diagnostic terms like *premenstrual syndrome*, which, as he says in the article, was previously called *premenstrual tension*. Greene cites Whitehead's 1934 study of woman pilots as a backdrop for some of his claims, but he offers a much more scientific account than Whitehead does of the reasons why female hormones might have caused these debilitating effects. The opening section of Greene's article discusses a list of symptoms that, just a few decades prior, were routinely affiliated with hysteria. He describes these symptoms as "neurological" and states that they include "epilepsy," "migraine," "severe headache," "vertigo," and "premenstrual *petit mal*." He then states that "*psychiatric* manifestations have been recognized for many years, especially that state of intolerable nervous tension or blackest depression which has broken up so many homes." He partially blames women themselves for not having managed this condition: "During the final week of the cycle many women experience great or small degrees of irritability which they may fail to control; depression which may lead to suicide; lethargy which may make it almost impossible for them to continue their work." He concludes this paragraph by referring to ancient beliefs about the lunar cycle's effect on human behavior, and he asks, "how many of us, using the word 'lunatic,' have suspected that we were referring to the premenstrual syndrome?" He makes specific mention of hormones when he says, "a shift in the oestrogen-progesterone ratio" might be responsible for water retention, and he then says, "oestradiol and progesterone are the only hormones proven to fluctuate regularly throughout the normal menstrual cycle."[52] He then discusses specific glands and specific portions of the endocrine system that might be responsible for these shifting conditions in women's bodies, as "few disorders in endocrinology are so simply explained." Specifically, he dismisses the hypothesis that premenstrual symptoms are caused simply by lowering levels of progesterone. Greene clearly had a more sophisticated and precise understanding of the menstrual cycle and all of the related disorders than his predecessor had. Nonetheless, many of the symptoms that were affiliated with hysteria are present in this article. And, in fact, the author mentions hysteria in a long list of possible side effects from the overdosage of progesterone.[53]

The enhanced vocabulary that was available to Greene becomes especially clear in the remainder of this article, which consists of case reports by physicians. These include descriptions of a wide array of symptoms—including "a mother of two young children" who "put her head in a gas oven," a thirty-four-year-old housewife who experienced severe asthma attacks only during her premenstrual phase, and a forty-nine-year-old married woman who experienced premenstrual migraines and ulcerative stomatitis every month. The only factor uniting these three cases was "the recurrence of symptoms always at the same time in each menstrual cycle." Furthermore, all these patients were treated with progesterone, which was reported to be a successful cure. As the physician acknowledges, though, there was no test for premenstrual syndrome, so the diagnosis "must depend on the intelligence of the patient, or her doctor."[54] In terms of connections to hysteria, the patient who put her head in a gas oven was temporarily confined to a "mental observation ward." After discharge, she was, according to this physician's report, "'rational and genuinely sorry for her action, which followed a family quarrel." The remainder of this passage narrates this patient's story from the husband's perspective: "Her husband readily confirmed that while she was normally happy and energetic, she always became depressed and irritable when her period was imminent, 'she suddenly changes, she looks for quarrels, I can't do anything to please her.' She was treated with 50 mg. progesterone and 2 c.c. mersalyl, and some sodium amytal to ensure a good night's sleep. The following day menstruation commenced and she was laughing again. She has since remained on progesterone and there have been no further attacks of acute depression."[55]

Several features are noteworthy in these patient case reports. First, the women's symptoms are reported and interpreted by males—either their husbands or the physician. The women have no agency to speak for themselves or interpret their own symptoms. Second, a collection of chemicals—including not only the hormone progesterone, but also the mercury-based diuretic mersalyl and the barbiturate derivative sodium amytal—replace the treatments that were previously recommended for hysteria. In the ancient texts, intercourse and pregnancy were often recommended to cure hysteria; in the mid-twentieth century, physicians recommended progesterone, which is the hormone that rises greatly during pregnancy. So even though these twentieth-century physicians did not recommend pregnancy as a cure, they did recommend its chemical substitute. Furthermore, as it was with the hysteria metaphor, in the ensuing analysis section of the article, a wide array of symptoms is discussed, and Greene, the physician, emphasizes that in some cases the woman's mental condition causes her to exaggerate the seriousness of these physical symptoms.

The next section of Greene's article presents a series of charts that visually represents patterns of individual patients' symptoms. The charts depict increases and decreases in blood pressure, albumin levels, and other symptoms, and they provide descriptions of each woman according to such characteristics as her age, occupation, and the number of children whom she has birthed. In some cases, the physician suggests that the woman's symptoms were restored to normal once she begins taking the right dosage of progesterone that would alleviate the symptoms without causing too many side effects. This is significant to note because it reinforces one important idea that crystallized as hormonal explanations replaced hysteria—namely, the idea that the goal of medical treatment was to make women constant so that they would be like men, who were presumed to be stable and constant in their bodily experiences and symptoms. One patient report, in particular, is important because it includes traces of the same language that was earlier used in reference to symptoms affiliated with hysteria. This patient is described as "a 40-year-old club hostess, who had been unemployed for six months owing to premenstrual epileptic fits and migraine." As with other patients, in this case the physician reports success in using progesterone: "No epileptic fits have occurred since commencing progesterone in September 1953 and she is now working happily as a shop assistant."[56] In other words, through use of the pregnancy hormone, this woman's health was restored, and furthermore, she became a productively employed citizen again. It is perhaps not coincidental that her earlier employment as a "club hostess" was replaced by the more respectable "shop assistant," which relates to the conflation of women's health problems with their morality that we saw frequently in experts' discourse on hysteria. Greene also repeatedly mentions throughout the article epileptic fits in a way that suggests that these fits had some relationship to premenstrual symptoms, but he never clearly explains the nature of that relationship.

Greene characterizes premenstrual syndrome as one of many "endocrine disorders." However, he mentions some external factors that could have exacerbated premenstrual syndrome, so it is interesting that he does not ascribe its cause exclusively to hormones. As he says, "at times of stress symptoms become unbearable and of increased severity, whilst when life flows along like a song the symptoms decrease or may pass unnoticed." Greene mentions and offers a clinically based explanation for the positive effects of progesterone, relating it to the feel-good effect of pregnancy and to the ancient practice of recommending pregnancy as a cure for hysteria. He reports that in his study, "62% were more energetic and free from symptoms during pregnancy," and he says that "it is probable that the extra progesterone produced by the placenta is responsible for the temporary remission." He reports, from the perspective of his patients who were prescribed progesterone as a treatment, that "they will sometimes return after a course of progesterone saying 'I feel wonderful—just as if I'm pregnant.'"[57]

After a long discussion of the positive effects of progesterone as a treatment for both premenstrual syndrome and toxemia, Greene concludes with a section about the "Social Consequences of Premenstrual Syndrome." This section begins with reference to hormones: "In view of the successful treatment with progesterone of premenstrual syndrome the social consequences merit serious consideration." As examples, he mentions the "high incidence of crimes" committed by premenstrual women, and he asks, "will a fine or prison sentence really cure the woman, who in a sudden uncontrolled fit of premenstrual irritability, throws a rolling pin at her husband or neighbor?" He then mentions Whitehead's 1934 study, which suggested that women pilots had a higher accident rate during menstruation. And, of course, he mentions women drivers who had "won the reputation for unpredictable action on the road." Green says, in his conclusion, that "during this phase of the menstrual cycle piece-time workers show lowered productivity, students become mentally duller, the suicide rate is higher and untold marital unhappiness and domestic discord have resulted from premenstrual outbursts of temper and irritability." Then, finally, he remarks that "the cost of progesterone therapy is high, but when this charge is weighed against the price in terms of human misery, suffering and injustice it is seen as a justifiable expense opening up a new vista of medicine."[58]

Conclusion

By thinking in terms of metaphor, we can appreciate the different forms of movement that were involved in the shift from hysteria to hormones; this movement involves not only changes in language but changes in the larger rhetorical and social context of medical diagnosis. As I have argued, mixing metaphors between the hysterical woman and the hormonal woman enabled four specific forms of rhetorical movement that can be identified in key medical texts since early twentieth century: argumentative shortcuts, purposeful confusion, expansion of empirical evidence, and expansion of expert vocabulary for female pathology. Coinciding with these four forms of rhetorical movement, we can also observe a larger movement that occurred in the rhetorical situation surrounding these topics. This larger movement consists of a shift from a situation in which a single disease, hysteria, was used to account for myriad symptoms that were related to women's health—even though the cause of this mysterious disease was never discovered—to a situation in which a single cause, hormones, was designated as the explanation for many of these same symptoms, although there was no disease to be diagnosed. It might seem like this shift would indicate that we have moved from darkness to light—as suggested in my analysis of the language used in the 1974 article that is quoted at the beginning of this chapter—because we now have a scientific explanation for a wide variety of "female problems" that had confounded physicians for centuries. Thus, we might hope that the symptoms would no longer be understood as pathological in the same way that they were when hysteria was the dominant metaphor. However, we can also draw on the aspect of metaphor that emphasizes movement and that has constant flux to note how this new metaphor—and the truths that it creates—have adapted themselves to old expectations so that women's bodies can still be understood as pathological, even when they are functioning normally. As rhetorical theorist Ken Baake observes, "such movement suggests that truth can never be fixed in language, but must be constantly renegotiated (constructed) as new metaphors appear."[59]

The rhetorical history that has been narrated in this chapter resembles, to some extent, the mixing of metaphors that has been noted in regard to other scientific subjects such as infertility discourses[60] and genetics,[61] but this chapter's analysis has revealed some distinguishing features that make the transition from hysteria to hormones an especially important shift in metaphors in the history of the production of endless numbers of new truths about women's health by (primarily male) experts. Of course, the fact that male experts are the ones who document what occurs in the bodies of female patients is nothing new; it has been a trend since ancient times. However, in the early twentieth-century texts that I have examined in this chapter, we see how hormonal explanations granted physicians greater confidence in their ability to account for these events in women's bodies. Whereas there was always much confusion in the days of hysteria about the source of these behaviors and symptoms, after hormones were discovered—and especially as this new discovery was implemented in scientific texts and treated as a matter of fact—physicians were much more confident in their assertions that hormones were the cause of these ill effects. After so many years of darkness and mystery, hormonal explanations shed a much-desired light on the mysteries of women's bodies, largely because they facilitated a more robust vocabulary for explaining these problems. Additionally, with the advent of chemical treatments came a new promise of something never before imaginable—namely, that women could be made to be more like men, and that their bodies could be tamed and controlled.

Continuing this book's efforts to speak back to the experts—both in medicine and in rhetoric—we again turn to Serres. Just as the female body caused trouble for medical experts, we can use the female body to trouble the concepts that rhetorical scholars have typically applied to texts and subjects that are as far removed as possible from the physical and material realities of the sexed body. Whereas metaphor has often been used in rhetoric to explain instances of language that make concepts clear to new audiences, or explain new and complex concepts in terms that are familiar to a given audience, Serres reminds us that lightness or illumination can only be seen because darkness exists alongside it: "One of the most beautiful things that our era is teaching us is to approach with light and simplicity the very complex things previously believed to be the result of chance, of noise, of chaos, in the ancient sense of the word. Hermes the messenger first brings light to texts and signs that are hermetic, that is, obscure. A message comes through while battling against the background noise. Likewise, Hermes traverses the noise, toward meaning."[62] The transition from metaphors of hysteria to metaphors of hormones, as explored in this chapter, is an ideal example for illustrating the claims that Serres makes about knowledge, light, obscurity, and noise in the above passage. One important lesson to be learned from studying the history of women's health is that every new scientific fact or discovery creates new areas of uncertainty or questions that need to be further investigated. In fact, the light that is cast by a new metaphor can only be seen if there is darkness to contrast with it, and we can only see areas of darkness when light is available. If we imagine those physicians who were "groping in the dark"[63] to understand women's bodies for so many centuries before the concept of hormones became available, then we are then better positioned to appreciate the light that hormones cast on these previously dark and mysterious topics.

Taking seriously Segal's assertion that diagnosis is a metaphor implies that we must look beyond specific categories or terms and consider more broadly how diagnosis itself, as a rhetorical act, is a metaphorical process. From this perspective, specific diagnostic terms like *hysteria*—and other diagnostic terms and the assumptions assigned to them that have come after it—are more than just names of specific diseases that experts agree to use in situations in which a particular group of symptoms manifests in a patient. Rather, as I have demonstrated in this chapter, hysteria was more than a disease, and it was replaced by hormones, which are also more than a contributing factor to disease. These are metaphors that shape how we think and talk about women themselves and the problems that they experience, both physically and mentally. As metaphors that have meanings in both scientific and popular discourse, they accomplish important rhetorical work. In particular, they uphold long-standing beliefs about women's inferiority. As I have suggested, hormones were the metaphorical vehicle that transported ancient ideas about women into the twentieth century, allowing these ideas to be communicated in ways that would meet the expectations of modern audiences who have come to demand scientific evidence and a scientific vocabulary to support the long-standing assumption that women were biologically inferior, which was supported by myth, religious beliefs, and superstition. The next chapter makes even clearer the significance of this metaphorical shift— and the new areas of darkness that were exposed by the light that it cast on female bodies—by revealing how the hormonal-woman metaphor has successfully transferred some key ideas to our understanding of key topics about women's health today.

7

This Is Your [Female] Brain on Hormones | Enthymeme in Contemporary Discourse

At a website called *Pregly*, a thread titled "It's not my fault I have pregnancy brain" appears in the "Pregnancy and fathers" discussion forum. Interestingly, despite the forum title, all the forum posts appear to be written by pregnant mothers, but these mothers emphasize how their "pregnancy brain" has impacted or inconvenienced their husbands. For example, the initial post in this thread says the following: "I keep forgetting that [my partner] has already told me things or that I've already asked him something and he's been getting really annoyed and nagging at me for not remembering something he had just told me. It's not my fault [sad-face emoticon inserted here] does anyone else's [partner] get frustrated with how forgetfull you are? Maybe it's because he's all hormonal. .haha." In response to this posting, several other pregnant women chime in to reassure this mother that they are all having similar experiences. An interesting refrain in these women's responses is that their behavior is not their fault, but rather the result of "pregnancy brain," which is explicitly or implicitly attributed to hormones. As one woman says in the forum, "omg exact same thing here!! My husband hates it!! He gets so annoyed!! But its not our fault!!!" Another chimes in to express her solidarity: "Im glad im not the only one!! I feel bad but its not like im doing it on purpose!"[1]

Is there scientific evidence to support the suggestion that women are made less mentally competent by pregnancy hormones? If not, what are some other frameworks in which we might try to understand this experience that some women report? Is it possible to treat women's self-reports of hormonal effects on the pregnant body as legitimate while still adopting a critical stance toward some of the precise language that is used to describe these effects, both in scientific texts and in women's own descriptions? And when scientific research produces evidence that substantiates or refutes a phenomenon such as pregnancy brain, who benefits from such knowledge?

This chapter addresses these questions in a manner that continues this book's mode of analysis—that is, by examining today's scientific texts as part of a continuation of the much older, even ancient, texts that characterized women's bodies as out of control, defective, and subject to change, a characterization that contrasted with men's bodies, which were assumed to be stable, manageable, predictable, and normal. As we have seen throughout this book, the notion that women's brains were influenced in distinct ways by their hormones or something inside the body has a long history. I argue in this chapter that the very fact that scientific experts continue to investigate a subject such as the effect of pregnancy hormones on women's brains perpetuates a long-standing assumption that women needed to concern themselves with their body-brain relationship, whereas for men this relationship was assumed to be seamless and unproblematic.

The key rhetorical concept in this chapter's analysis is the enthymeme. Of course, traditional definitions cast the enthymeme as an abbreviated syllogism, often based on Aristotle's description: "The enthymeme must consist of few propositions, fewer often than those which make up the normal syllogism. For if any of these propositions is a familiar fact, there is no need even to mention it; the hearer adds it himself."[2] However, contemporary rhetorical theorists have rejected this narrow definition. M. F. Burnyeat, in particular, denies that Aristotle even intended the term *enthymeme* to denote a syllogism with a missing premise: "It is a good enthymeme, not an enthymeme as such, which omits to formulate premises that the audience can supply for themselves, where a 'good' enthymeme is to be understood, again by reference to the function of rhetoric, as one that is effective with an audience of limited mental capacity."[3] As this passage indicates, the term *enthymeme* is used more loosely in contemporary rhetorical theory than in classical rhetoric. And for many contemporary theorists, enthymemes can work for any audience, not just those of "limited mental capacity."

The previous chapter emphasized how metaphor facilitates the movement of ideas through time. In this chapter, we will see how enthymeme also facilitates movement, but the emphasis is on the movement of ideas across physical, digital, and geographic space. The movement that is facilitated by enthymeme, I will argue in this chapter, is a movement within the minds and bodies of audiences at a given time and place, but it is also a movement that can spread rapidly beyond its point of origin, such as when a scientific study receives much media attention and then feeds into the frenzy of popular beliefs that can surround a topic such as pregnancy brain. If metaphor facilitates movement by presenting something—often unfamiliar or abstract—in terms of something else that is usually more familiar or concrete, enthymeme moves its audience by letting them stay in that familiar place and remain comfortable even as they are presented with new ideas that might otherwise seem disagreeable, far-fetched, or unsettling. An enthymeme is effective for a particular audience in a particular time and place because it fits into a well-worn groove in the hearts and minds of its audience.

Jeffrey Walker invokes Isocrates's definition to explain how enthymeme facilitates this kind of gentle, yet forceful, rhetorical movement. In Walker's words, effective enthymemes "will seize the kairos of the moment to move the audience to a decisive recognition that is or seems 'lofty and original,' while at the same time 'cutting off' or shifting into the background other possible recognitions that may be latent in the buildup." Along these lines, Walker draws a parallel to Kenneth Burke's concept of identification, noting that an effective enthymeme will "generate in its audience a passional identification with or adherence to a particular stance, and that (ideally) will strike the audience as an 'abrupt' and decisive flash of insight."[4]

Matthew Jackson's reconceptualization of the postmodern enthymeme adds to our understanding of the movement that metaphor facilitates, with an emphasis on the audience, speaker, text, and rhetorical situation that define the spaces that enthymemes enable ideas to traverse. As Jackson says, "a reconceptualization of the enthymeme with a postmodern sensibility would allow for more than one premise to be supplied by individuals as part of a fiction of the audience in order to make multiple meanings. Furthermore, these meanings can be tentative and dynamic. . . . In the postmodern enthymeme, the speaker, text, and audience are fragmented, shifting, dynamic, multiple, and not always real (in a positivistic tangible way)."[5] In short, if we adopt the broadest definition, enthymeme is any aspect of an argument whose omission makes the argument seem compelling or intriguing or novel and, at the same time, makes its audience feel comfortable or satisfied. An enthymeme might shake us up a bit, or cause us to view some aspect of our world a little differently, but it does not threaten our core beliefs or values in any significant way. An enthymeme is effective in situations in which the audience is inclined—for any number of reasons—to accept the argument without the piece, or pieces, that are left unstated. Thus, an enthymeme can be any element in an argument that is not strictly rational but is powerful despite this lack of rationality (or perhaps because of it). Because an enthymeme depends on shared assumptions among its audience members, we can often discover an enthymeme by looking critically at an argument from a perspective that rejects the unstated beliefs on which the enthymeme depends. As we see this form of rhetorical movement enacted in contemporary science of female hormones, we add to the contemporary understanding of this rhetorical concept by emphasizing that enthymeme involves the omission not of just any portion of a scientific argument, but of key pieces of a scientific argument that might cause audiences to rethink fundamental beliefs that they have long held dear.

Taking these contemporary theories of the enthymeme as a starting place, this chapter explores multiple strands of recent research that investigate the hormone-brain relationship in women. For each strand of research, I expose what has been omitted and how this omission serves as an enthymeme that causes the argument to resonate with specific expert and public audiences. First, I examine recent research that attempts to substantiate women's anecdotal reports of "pregnancy brain." When we look closely at scientific and popular arguments about the effects of pregnancy on women's brains, we see that these arguments are often constructed in ways that omit scientific evidence in order to support claims about hormones as the cause of cognitive diminishment. Second, I consider recent research about hormones and women's mental health. In this research, the question is less about cognitive impairment than it is about depression, anxiety, and mood disorders. And the focus here is not on one distinct period of a woman's life, but on all the flux and change in hormones that occur throughout the woman's life, with the suggestion that this flux and change mean that even women with "normal" menstrual cycles are more susceptible than men are to mental-health problems such as depression. The function of enthymeme in this body of research becomes most apparent when we scrutinize the implicit value judgments that experts make about positive and negative responses to stressful events and the assumptions that guide research about male versus female hormones as these relate to the aspects of stress being studied; omitting explicit mention of or elaboration about the reasons that underlie these value judgments makes the arguments more persuasive because it allows these value judgments to resonate quietly with popular beliefs about appropriate and inappropriate responses to stressful events and does not require the audience to question these popular beliefs. Third, I examine recent literature in which researchers use evolutionary narratives to emphasize the possible benefits of pregnancy hormones for the brain, such as neuroplasticity, and I consider how this emphasis leads to more nuanced conclusions about ways that some sections of the brain have reportedly benefitted from pregnancy hormones, while others are harmed. The function of enthymeme in this body of research is most apparent when we consider the distinctions that researchers make among the various kinds of cognitive skills that have supposedly been enhanced by different kinds of female hormones; close examination reveals that experts' judgments about the value of these skills reflect long-standing, yet unsubstantiated, assumptions about the skills that women need in order to succeed in the domestic sphere in contrast to the workplace, and leaving these assumptions unstated makes the arguments more, not less, persuasive.

Pregnancy Brain: Fact or Fiction?

Two important review articles about the subject of so-called "pregnancy brain" were published in the first decade of the twenty-first century. The first was "Motherhood and Memory: A Review," by Matthew Brett and Sallie Baxendale, published in 2001 in the journal *Psychoneuroendocrinology*. The second was "A Review of the Impact of Pregnancy on Memory Function," by Julie D. Henry and Peter G. Rendell, published in 2007 in the *Journal of Clinical and Experimental Neuropsychology*. These two articles were highly visible: Brett and Baxendale's review currently has ninety-nine citations in Google Scholar, and Henry and Rendell's has sixty-four citations, and they have both received popular media coverage as well. Looking closely at the arguments that are presented in these two articles reveals one type of omission that serves an enthymematic function in the body of literature that is evaluated in this chapter: the omission of evidence that supports the claim that hormones were the cause of the cognitive deficits that pregnancy reportedly caused.

Because "pregnancy brain" is an experience that many mothers self-report, it is common for authors of scientific articles about the subject to begin with a reference to anecdotal reports and then to position their study as an effort to find definitive evidence that either supports or refutes the validity of these mothers' own reports. Feminist scholar Nicole Emily Hurt describes this rhetorical pattern as the "baby brain dilemma," which she says emerges from "the inconsistency between women's reports of memory loss during pregnancy or the postpartum and science's inability to validate this experience." As Hurt says, in creating the exigency for their research, these authors actually "pit women's experiences against science."[6] Exemplifying this rhetorical pattern, Brett and Baxendale begin their review article by addressing what they call "subjective reports" of the cognitive decline that many women report during and immediately after pregnancy. Interestingly, many of the popular articles that they cite are written by women professionals (mostly physicians) who describe in great detail the cognitive impairments that they claim to have experienced during pregnancy. The symptoms that these women report include not only "forgetfulness and a poor memory," but also "distractibility, weak concentration, word finding difficulties, poor co-ordination, and general cognitive slowing." Brett and Baxendale note that we need better scientific understanding because if we have only subjective reports from women themselves, we might fail to recognize fully the cognitive impairment that these women experienced: "Complaints of 'memory difficulties' may simply reflect an overall deterioration in cognitive function."[7]

After this brief discussion of "subjective reports," Brett and Baxendale turn to "objective reports" of memory impairment, structuring their discussion around previous research about two different kinds of memory: implicit (nondeclarative) and explicit (declarative). One of the first studies that they discuss is a 1999 study that suggests that maternal rats, and even foster-mother rats with pups, performed better and more quickly than nonmaternal rats in a food maze. They surmise that because this improvement in "implicit memory" was the same for foster mothers as biological mothers, the effect was not necessarily a result of hormones: "It is possible that the presense of pups for both the maternal and foster-rats increased their anxiety/hunger levels and thus improved their performance."[8] Although evidence such as this, which suggests a cognitive benefit of pregnancy and mothering, could have been used to support a more nuanced understanding of a phenomenon such as pregnancy brain, the authors instead use it to set the stage for the argument that they clearly want to emerge as the most salient conclusion of their review article: that pregnancy brain is real, and that hormones are the culprit. Thus, when they reach the summary section of their article, they acknowledge that "there is conflicting evidence regarding the impact of pregnancy on implicit memory skills." However, despite this lack of conclusive evidence, and even though they cite some studies that suggest the cognitive benefits of pregnancy, the authors repeatedly emphasize the likelihood that hormones were the culprit of the memory impairment that many mothers self-reported. In fact, the authors seem so sure that this memory impairment is real that they skip many logical steps and arrive at the assertion that medical experts needed to agree on terms to diagnose this condition: "Although there have been calls to recognise 'maternal amnesia' or 'benign encephalopathy of pregnancy' as a syndrome, there is as yet, no clear picture of the core symptoms." As they proceed toward providing this clear picture, Brett and Baxendale suggest that "hormonal changes" are a possible cause of these cognitive deficits, but they also state that "the cause, nature and course of 'maternal amnesia' remain uncertain."[9]

Despite this uncertainty, which the authors repeatedly acknowledge, the authors note in a subsequent section of the article that "many hormones have altered levels in pregnancy . . . and four groups of these hormones are known to influence brain structures involved in memory."[10] In this paragraph, the authors provide no citations for this assertion that hormones influenced brain structures, but they proceed nonetheless to discuss each of the four hormones individually, including estrogens, progesterone, glucocorticoids, and oxytocin. Although they provide in-depth coverage of existing research about the effects of these hormones on "memory structures in the brain," the evidence that they present on all four of these hormones is ultimately deemed, once again, to be inconclusive. The following paragraph, which summarizes the conflicting evidence that the authors report regarding estrogen's effects on memory, is typical of the stance that these authors adopt toward the available evidence pertaining to each of the four hormone types: "However the effect of very high levels of oestrogen on memory skills is unclear, and in particular, there is little evidence to suggest that very high oestrogen levels have a beneficial effect on memory. The net effect of oestrogens seems to be excitatory, perhaps predisposing to excitotoxic damage, and it is possible that oestrogen metabolism is toxic to cells in some circumstances." In this paragraph, the authors suggest that even though "oestrogens seem to play a role in maintaining or even augmenting memory," the high levels that occurred during pregnancy might cause this effect to be reversed.[11] Thus, even though the authors acknowledge that there is no definitive research that substantiates the claim that estrogen affected memory during human pregnancy, they resolve this apparent contradiction by suggesting how this inconclusive evidence can be aligned with anecdotal evidence of reduced memory function during pregnancy. The most conclusive research that the authors cite is a group of studies that demonstrate that women struggled with verbal recall after bilateral ovariectomy and after menopause. Both of these are situations in which loss of estrogen is presumed to have caused the memory loss. Their other claims, which they intend to use in their explanations about how increased estrogen levels could also have led to memory loss, are based purely on speculation and include no supporting evidence.

Despite the uncertainty that the authors acknowledge about the available scientific evidence, in the next section they jump ahead to provide a "clear picture of the core symptoms" of "maternal amnesia" that they called for earlier in the article. Specifically, they identify two different kinds of memory loss that could have occurred during pregnancy or, in their words, "two syndromes of memory loss associated with pregnancy." The first was "gestational memory impairment" (GMI), and the second was "prolonged post-partum memory impairment" (PPMI). GMI, the authors write, occurred in about 80 percent of pregnancies, usually during the second and third trimesters. It was "characterized by a high level of subjective memory complaints and an objective impairment on explicit memory tasks." This syndrome "appears to resolve soon after childbirth, although there is little direct evidence to support this." PPMI, by contrast, "seems to date from childbirth, rather than pregnancy, and to be long lasting." PPMI "appears to be much rarer" than GMI, but "the lack of longitudinal studies of memory following childbirth means that the true prevalence of such complaints is unknown."[12] So, in short, even though Brett and Baxendale do not cite any conclusive evidence, they name two syndromes, both surrounded by a great deal of uncertainty.

In the final paragraph of this section, the authors summarize their findings. Interestingly, they seem to emphasize, in much bolder terms with less hedging, the conclusions that were articulated in the previous sections—that is, that hormones were the main culprit in the memory loss that came with pregnancy. They also claim that the memory loss that some women experienced during pregnancy resembled the hippocampal damage experienced by people with brain damage, which suggests to the authors that pregnancy hormones might have caused hippocampal damage: "In our review, we have concentrated on the hormonal changes of pregnancy, because several hormones that are elevated during pregnancy have specific effects on memory. We have also suggested that the patterns of memory loss both during and after pregnancy have features of hippocampal dysfunction. In our cases, the pattern of memory loss was consistent with reduced hippocampal function, with impaired recall, but not recognition, and no impairment in non-memory tasks. We therefore propose that both

GMI and PPMI may be partly the result of hormonal action on the hippocampus."[13] As they reach the end of the article, Brett and Baxendale continue to acknowledge the lack of trustworthy scientific evidence to corroborate the phenomenon of baby brain. As they say, "memory loss following pregnancy and childbirth has not been well researched, and whilst anecdotal evidence abounds, rigorous research designs have yet to be applied." Thus, they conclude the review article by calling for studies that will provide empirical evidence to "establish the existence, prevalence, and clinical significance of GMI and PPMI." Such studies will need to "establish the extent of both the organic and psychological bases of the complaints."[14]

Brett and Baxendale's review article illustrates how enthymeme operates through the omission of key information. Because expert and nonexpert readers already share so many assumptions about the detrimental effects of pregnancy hormones on the brain, there is no need for the authors to provide definitive scientific evidence to support claims of this nature. In fact, omitting such evidence is an effective rhetorical move because, as is the case for all good enthymemes, this omission allows the arguments to be more economical and concise. Furthermore, the gaps that remain in the available evidence that the authors cite, along with the numerous anecdotal reports from women who had experienced so-called pregnancy brain, fuel the authors' repeated statements about the need for more research; this enthymeme keeps the argument moving.

Henry and Rendell's 2007 review article can be seen as a continuation of this same movement. In their opening paragraphs, the authors cite Brett and Baxendale's 2001 review article, among others, as examples of studies that report "robust findings of perceived deficits" in cognitive function during pregnancy.[15] They begin the article by echoing Brett and Baxendale's assertion that women have often self-reported memory loss during pregnancy, and using this to suggest a need for scientific evidence to substantiate these self-reports. Henry and Rendell's meta-analysis of fourteen studies is an attempt to extrapolate more certain knowledge from the "relatively large body of empirical evidence" that "focused on quantifying the nature and the magnitude of pregnancy-related memory change."[16]

The authors emphasize the need for further "exploration of the etiologies and functional consequences of pregnancy-related memory difficulties."[17] In other words, they suggest that the cause of these memory difficulties is not yet known. However, other places in the article (and in subsequent articles that cite this article), the authors are quick to assign hormones as at least one of the causes of this memory loss, using rhetorical moves that are similar to the ones that were apparent in Brett and Baxendale's article. For instance, Henry and Rendell state toward the end of the article that "whilst further research is needed to delineate the specific factors that may lead to memory failures in pregnancy and postpartum, it is of note that both free recall and the executive component of WM impose particular demands upon executive cognitive control processes."[18] The authors then cite several other studies that attribute different causes to the memory deficits that were experienced by pregnant and postpartum women. One such study attributes these deficits to "the complex hormonal changes that take place during pregnancy and birth," but others mention different causes, all quite vague.[19] The article ends by asserting the need for more research about the etiologies of memory deficits that occurred during pregnancy and the postpartum period. Thus, we see in both of these review articles a similar pattern: the authors begin by referring to anecdotal reports of women's experiences with pregnancy brain, and they use these reports to justify the need for research that will lead to more definitive evidence. Because they fail to find this evidence themselves, the authors return to the anecdotal reports as a justification of the need for further research. The circular nature of this argument can be seen as another manifestation of the kind of rhetorical movement that enthymeme facilitates; the omission or gap is created by the failure of the authors ever to discover the kind of definitive evidence that they are looking for. Even though researchers never find this evidence, however, they move forward with their arguments, which seem to become more rather than less compelling because of this gap.

Hurt's feminist critique of Henry and Rendell's review article exposes another omission in these authors' arguments. As Hurt says, Henry and Rendell do not account for demographic differences or different stages in pregnancy and the postpartum period. Rather, "pregnancy and the postpartum period . . . are biological conditions" in this study. The only common factor uniting the fourteen studies that are included in this meta-analysis is that each study compares pregnant and/or postpartum women to a control group. As Hurt says, this approach "only makes sense if pregnancy is assumed to be a homogeneous condition that women from various corners of the world and throughout different decades experience in roughly the same way."[20] In other words, the main feature that comes to define women as a group is their pregnancy hormones— this is what makes them who they are, despite social, cultural, racial, or economic differences.

The biggest problem with this situation, Hurt concludes, is that it presents baby brain as a completely incurable, biological fact that women just had to deal with. She concludes with a discussion of possible alternatives in which, we might acknowledge, this phenomenon exists to some extent, but it can be seen as a cultural phenomenon that could be addressed through changes in the social and political structures that govern our existence: "Instead of linking women's concerns about their cognitive abilities to their hormones, we ought to focus on the structures in our society, such as household and workplace expectations, that make motherhood a difficult task."[21]

Through detailed analysis of the authors' statistical reporting, Hurt points out that pregnant and postpartum women actually outperformed the control groups in three areas: recognition, implicit memory, and short-term memory. Furthermore, Hurt notes that even when the authors report statistically significant differences, these are "small in terms of effect size." Hurt ultimately argues that "at most, the pregnant women's statuses as being pregnant explains only about 5% of the variation in scores on memory tests," and "based on these numbers, Henry and Rendell's study could have been interpreted as providing little evidence to solve the baby brain dilemma." Of course, this was not the message that was communicated to public audiences. Rather, as Hurt says, "the media presented Henry and Rendell's study as having finally solved the baby brain dilemma, thereby granting scientific legitimacy to a dangerous myth."[22] She also notes the manner in which Henry and Rendell elide the word *statistical* from their use of the term *significance*, suggesting that this rhetorical move is confusing both to the expert audiences who would likely read the scientific article and to the popular audiences who would read about these findings in the mass media.

Building on Hurt's feminist critique of Henry and Rendell's study, we can see in their article—and news reports that followed—how hormones have come to serve a particular rhetorical function in the movement of arguments about pregnancy brain as these arguments circulate among both expert and public audiences. If, as J. Scott Blake explains, an enthymeme "draws its rhetorical force from the *movement* of a chain of premises and its surrounding complex of proofs,"[23] then we start to see how scientific articles and news reports can reinforce each other in a perpetual spinning and remixing of premises and proofs, none of which necessarily add up to a valid logical syllogism. However, through their cumulative force, these reports produce an argument that is powerful because it builds on its own self-perpetuating momentum. Scott adds that "an enthymeme not only plays on an audience's assumptions, but can also help shape those assumptions."[24]

Other examples reinforce that when we are dealing with a topic such as pregnancy brain, the movement that enthymeme facilitates between expert and public spheres can be especially quick and powerful. For example, a 1997 article in *New Scientist* made the following report about the results of an MRI study that supposedly showed shrinkage in the brain volume of ten healthy women toward the end of pregnancy: "Some pregnant women complain that their brains have gone on leave—they find it difficult to concentrate, and have poor memories. Now, an explanation may be at hand: a group of anaesthetists and radiologists believe that women's brains shrink during late pregnancy and take up to six months to regain their full size."[25] The full findings of this study were reported a few years later in the *American Journal of Neuroradiology*. Not surprisingly, the full findings included a much more nuanced level of detail, and the authors' narrative in the scientific article focused on differences between the brain sizes of healthy pregnant women and those who were diagnosed with preeclampsia. The authors even ended the scientific article with an acknowledgment of uncertainty about the meaning of this decrease in brain volume: "The precise mechanism and physiologic importance of the changes is not known at the present time."[26]

Both of these articles—the shorter *New Scientist* report and the full-length research article—received much attention in the popular media. As noted at the beginning of Brett and Baxendale's review article, Pete Moore's *New Scientist* article caused other news media outlets to report the study about brain shrinkage, and this scientific news spread rapidly in a way that reinforced popular notions about "baby brain" and other cognitive deficits that mothers supposedly experienced during pregnancy.[27] Even as recently as 2016, a popular blogger linked to the research article to support her assertion that "your brain shrinks . . . literally" during pregnancy.[28] Through examples like these, we can see how enthymeme exists not only at a single moment or element in an otherwise rational argument, but as a component of a complex "rhetorical ecology"[29] in which ideas that seem compelling to an audience get reinforced, again and again, by circulating through different forums, both public and expert. When we consider the different forms of omission that work together in examples such as these to produce powerful sets of arguments, it becomes clear that enthymemes can operate on many different levels. If all the nuance that is presented in scientific articles were included in all of the other sources that help new findings about pregnancy brain circulate rapidly in the public sphere, few readers would take the time to sort through the conflicting data. As a result, the ideas would likely neither circulate nearly as rapidly nor cause so much buzz among various audiences.

If this phenomenon were limited only to distinct phases in a woman's life, such as pregnancy, that would be problematic in and of itself. But, as we will see in the next section, there is another body of research that seeks to investigate the effects of hormones on women even during a perfectly normal menstrual cycle. Examining this body of research exposes additional forms of omission that can constitute enthymeme in arguments about female hormones and brains.

Effects of a Normal Menstrual Cycle on the Female Brain

One of the features that contemporary hormonal discourses preserve from the past is that these contemporary discourses leave open the possibility that all women are pathological because of their female hormones (not because of a specific condition or diagnosis that was attributed to abnormalities in the endocrine system, and not because of a specific event such as pregnancy)—a theme that continues from the previous chapter. Recent research continues the efforts of scientists from previous eras to explain the fluctuation, change, and apparent instability of women's mental capacity, relative to men's, even during a normal menstrual cycle. Thus, in the studies examined in this section, we will see another way in which enthymeme can function in scientific arguments about women and hormones. As with the literature on pregnancy brain, in these arguments, enthymeme is grounded in forms of omission that make the argument more rather than less powerful. In the literature discussed in this section, however, the key piece of the argument that is omitted is an explicit articulation of the value judgments that researchers make regarding positive and negative reactions to stressful events.

A 2015 article titled "Estradiol Levels Modulate Brain Activity and Negative Responses to Psychosocial Stress across the Menstrual Cycle" is a recent example in which this omission appears. In this article, the authors report the results of a study that enrolled twenty-eight women with normal menstrual cycles and that used the Montreal Imaging Stress Task (MIST) to test their brain activity in response to "psychosocial stress" at different phases of the menstrual cycle. In particular, the researchers looked for differences between the low-estradiol and high-estradiol phases of the cycle.

As for the study methods, the introduction claims that the authors are interested in a subset of psychosocial stress called "social-evaluative threat." To test women's responses to this type of stress, "participants were recruited and told that they would be participating in a study to examine the effects of menstrual cycle hormone changes on mathematical ability in women." The particular stress task, the MIST, is later described as "produc[ing] moderate psychosocial stress through a combination of motivated performance and social-evaluative threat." Subjects were given an arithmetic task and told that they needed to achieve between 80 and 90 percent correct answers, but the test included an algorithm that kept making the questions more difficult to ensure that no participant would score higher than between 40 and 50 percent. The social-evaluative threat was imposed by experiment staff, who followed a script that involved periodically entering the exam room to tell participants that they were not scoring well enough. As the authors say, "the interaction of the participants and the experimenters was structured to generate psychosocial stress."[30] In short, women subjects were being tested on their ability to thrive—that is, to remain calm and avoid experiencing high distress levels—in a situation that was designed to ensure that they would fail at a difficult intellectual task. Even though they were informed that the test was designed to evaluate their mathematical ability, these subjects were not actually being tested on their ability to complete these arithmetic problems, but rather on their ability to overcome obstacles purposely designed to encourage them to fail.

The researchers found in this study that higher levels of estradiol correlated with more "positive" responses to psychosocial stress. This finding corresponded with their hypothesis: "We hypothesized that high circulating estrogen would reduce the effect of acute stress on brain activity and be associated with less subjective distress following the stress-inducing task."[31] In line with this hypothesis, these researchers found that women with higher estradiol levels had lower distress scores after the stress task, and the researchers judged this to be a more positive response to stress because it produced fewer negative effects on the women's moods. Two different measures were used to arrive at this conclusion. One was a subjective measure, asking participants to self-report their distress levels, and the other was a more objective measure, which was based on the detection of brain activity through an MRI scan. During "low estradiol phases of the menstrual cycle," the authors conclude, the low estrogen levels led to "periods of increased vulnerability to psychosocial stress."[32] In short, women with lower estradiol levels had higher distress scores and a "more negative mood" after the MIST was administered.

The enthymeme here arises from a purposeful omission of any elaboration on the principles or values that guide these researchers' assessments of what counts as positive and negative responses to psychosocial stress. This omission becomes apparent in language such as the following: "Women with higher distress had lower left hippocampal activity during the MIST, more negative mood following the MIST, and lower estradiol levels."[33] The phrase "negative mood" suggests an underlying assumption that these women who participated in the study should have had a positive mood after enduring a situation in which they were bombarded with difficult arithmetic problems and study personnel who acted intentionally to make the environment more stressful. Regardless of the methods that were used to measure such responses to stress, we have to look closely at the value judgments that underlie this experimental process. It is clearly expected that the women who had a more intense response to this stressful situation—whether that response is measured "subjectively" through self-reporting or measured "objectively" through an MRI examination of brain activity—are judged negatively in relation to the women who had a less intense response. Thus, the researchers begin with an expectation that when faced with difficulties imposed from the outside, there was a predefined way in which successful or healthy women should have responded. The ideal subject should have exhibited steady physiological responses, exemplified in this study by minimal disruption in the "normal" amount of hippocampal activity, and she should have subjectively reported, when asked on a questionnaire, minimal amounts of those emotions that are characterized as negative.

In seeking to understand the enthymematic component of this argument, some important points of contrast can be observed between this research and pregnancy-brain research. As in pregnancy-brain research, hormones are assumed in this study to be the cause of the "negative" moods that women reported during the low-estradiol phases of their menstrual cycles, and the authors provide little direct evidence to support this cause-effect relationship, so this same enthymematic function is present in the arithmetic study. But in this case, we also see how any given scientific argument can include multiple enthymemes, with proofs and claims intermingling with each other in unpredictable ways. As Walker says, "in a large and complex argument," we can expect to see not just one enthymeme, but also "a progression from enthymeme to enthymeme to enthymeme, building up an accumulated fund of value-laden, emotively significant ideas (oppositions, liaisons, etc.) that are variously brought to bear, forcefully and memorably, in the rhetor's final enthymematic turns."[34]

The net effect in this case is not only that hormones are assumed to be the cause of these effects, but also that researchers can employ a powerful set of unstated value judgments about the negative and positive qualities of the cognitive effects that are in question. These value judgments can only be understood to make sense if we accept a whole set of cultural and social norms that have nothing to do with scientific proof but everything to do with the habits and practices that have come to govern our assessments of different individuals' responses to stressful situations.

To see more clearly these value judgements that lie underneath the surface, we need to look at the instruments that these investigators used in this study to measure subjective responses to stressful events. One of these is the "stress/arousal adjective checklist." For this measurement, subjects were asked to rank whether each word in a list of adjectives (including adjectives such as *calm, worried, distressed, uptight*, etc.) described their state at a given time. They could choose from among four options: definitely yes, slightly yes, not sure/don't understand, or definitely not. The responses were dichotomized, with each response classified as a positive or negative response, depending on whether the adjective was believed to describe a healthy or unhealthy response to stress. In all of these distinctions, there is clearly an inherent value judgment about which responses are positive and which are negative, and leaving out an explicit articulation of the reasons for this value judgment serves as an enthymeme that enables the argument to resonate with popular beliefs about normative female responses to stress. This is important because leaving out this piece of the argument enables us to avoid asking some important questions about implications for the real world beyond this study. For instance, if a subject described herself as "calm" after being bombarded with this stressful situation, what might that mean for this person's likely behavior in the workplace? It might mean that this person is willing to put up with terrible treatment and more likely to accept mistreatment, for example. Although that might make this woman a better employee, and a more gender-conforming individual, it might not mean that she is a better, or healthier, person overall or that she is happier and more successful in the workplace. Another woman might have reported a higher distress level after being bombarded with this terrible and unfair situation, but maybe that means that this woman would be more likely, for instance, to stand up to a bully or sexual harasser. Maybe this woman would be more likely to risk her own status and take a stand against unfair practices that she sees her colleagues experiencing, or maybe she would be more likely to look for another job instead of continuing to endure unfair treatment with an outwardly calm demeanor. The point is that there are unstated assumptions about the values that these researchers used to distinguish positive, healthy responses to stress from negative, unhealthy responses. Leaving these assumptions unstated makes the argument more powerful because it is a simple argument that also resonates with widely shared assumptions about gender difference. Basically, a study like this gives us scientific evidence to support a set of beliefs about the proper behavior of women—or, more precisely, proper female responses to stressful events. These beliefs have long circulated in our culture and have gained wide acceptance, but a study like this can strengthen such beliefs by providing the evidence that is needed to present such beliefs as scientific facts.

Feminist psychologist Jane M. Ussher claims this is a widespread tendency in much of the psychology research that attempts to explain why women experience higher rates of depression and anxiety. These hidden value judgments reflect what Ussher calls a "biomedical approach" to women's mental health, which assigns the blame to hormones, or "pathologizes femininity." As Ussher says, "gender role stereotypes used by medical practitioners and gender bias in psychometric instruments that categorize normative aspects of feminine behavior (such as crying or loss of interest in sex) as 'symptoms' have been deemed to result in medical practitioners diagnosing depression in women at higher rates than men." Premenstrual dysphoric disorder is one powerful example. As Ussher asserts, "while premenstrual change is a normal part of women's experience, feminists have argued that these changes are only positioned as 'PMDD' because of hegemonic cultural constructions of the premenstrual phase of the cycle as negative and debilitating, which impact upon women's appraisal and negotiation of premenstrual changes in mood or behavior." This does not happen in other cultures, Ussher suggests. She further asserts, "the gendered nature of this medicalization results in an insidious creeping of pathologization into women's lives." She argues, furthermore, that even presumably normal events in women's social and biological lives, including marriage, menopause, employment are constructed as "psychiatric illnesses that warrant SSRI medication."[35]

In the Albert, Pruessner, and Newhouse study about psychosocial stress, the phenomenon that Ussher describes is exposed by scrutinizing the instruments that the authors used to measure so-called subjective responses to stress. The overarching framework of the study is informed by the use of MIST data. MIST is a widely used instrument for testing both men's and women's responses to stress in all kinds of conditions, so the choice of this instrument is not in itself a signal of misogynistic intentions. However, some troubling gender disparities in this study's implementation of the MIST instrument can be highlighted by looking at another recent study that was designed to test the fluctuation of testosterone levels in both male and female subjects after they underwent the MIST testing. This 2015 study, titled "Significant Association of Testosterone and Prefrontal Stress Response Is Specific for Males,"[36] not only enrolled a different population that included both males and females, but it also had a different design. Rather than measuring subjects' hormone levels before they began the test, this study measured testosterone levels in males and females after the test. The key finding was an increase in testosterone levels in the male subjects after the stress test, but not in females, and not surprisingly, the rise in testosterone levels is judged by the authors as a positive response to stress. So there is a reverse cause-effect relationship at play in this implementation of the MIST: instead of conceiving of *female* hormones as instigators of stress response, in this study the researchers looked at *male* hormone levels as the effects of the stress response. This contrast between the goals and design of these two studies highlights the gender-specific assumptions that are left unstated and, therefore, can serve as powerful sources of enthymeme in the Albert, Pruessner, and Newhouse study.

There are other forms of omission in the Albert, Pruessner, and Newhouse study as well. Echoing a feature of pregnancy-brain research that Hurt observes in her feminist critique of this type of research, Albert, Pruessner, and Newhouse define the problem that motivates their study in a way that excludes the consideration of the complex set of social, economic, and institutional factors that likely contribute to the higher rates of depression and anxiety that women reported experiencing. This omission becomes clear in the first sentence of the abstract, in which the authors explain how their study proposes to advance current scientific understanding: "Although ovarian hormones are thought to have a potential role in the well-known sex difference in mood and anxiety disorders, the mechanisms through which ovarian hormone changes contribute to stress regulation are not well understood."[37] As Ussher argues, the statistics about higher rates of depression in women that have often been cited to justify research of this nature should not be taken at face value. Ussher observes that in biomedical explanations for women's higher rates of depression, female reproductive hormones "in particular, oestrogen, linked to premenstrual, post-natal and menopausal stages of the reproductive lifecycle"—have been assigned as the culprit, which Ussher refers to as the "raging hormones" approach.[38] Ussher cites a number of studies that suggest alternative explanations that acknowledge social and cultural factors. Ussher's discussion provides a much more nuanced picture of the statistics that suggest that women suffer higher rates of depression and mood disorders, and a useful counternarrative to Albert, Pruessner, and Newhouse's emphasis on hormones as the main contributing factor to women's higher rates of mental-health problems. Despite all of this uncertainty that is pointed out in the literature of feminist psychology, Albert, Pruessner, and Newhouse strictly adhere to what Ussher calls the "realist biomedical model," which entails taking statistics at face value to suggest that it is a well-established truth that women suffer higher rates of depression and other related mental-health conditions. It is also, of course, interesting to see these researchers going further by taking advantage of the more sophisticated brain imaging technologies that are currently available to provide scientific evidence for a phenomenon that has long been suspected but never documented or scientifically explained.

There is other notable language in the way that the study is explained in its opening paragraph: "Psychosocial stress may be especially important in the etiology of mood disorders in women as the depressogenic effects of life stressors are reportedly greater in women than men ... even when there is no difference in the number of stressful life events or in the subjective perception of these events."[39] In this passage, the authors essentially say that women have been demonstrated to have a more severe response to life's stresses than men do. This response would be a difficult phenomenon to measure, and even though the authors say that they are separating this measurement from the number of stressful events, it is quite difficult to tease those apart. Despite these difficulties, however, the authors find a way to support these claims with empirical measurements, and they proceed with the study, as I have outlined above. As we will see in the next section, even in some of the recent research that claims to find more positive effects of hormones on the female brain, this tendency to obscure the social and cultural factors that underlie scientific assessments about women's cognition and behavior is still a persistent means of generating enthymemes that allow these arguments to resonate with expert and popular audiences alike.

Seemingly Positive Cognitive Effects in the Most Recent Literature

One shocking suggestion that appears in some recent research about female brains and hormones is the idea that the female brain is more complicated than the male brain is because it has to adapt to the hormonal flux that comes with the female menstrual cycle. For example, a 2013 review article includes the following language: "The female brain must have mechanisms in place to cope with the monthly fluctuations of sex steroids, many of which are neuroactive, and it is probably only when such adaptive mechanisms are disturbed that psychiatric diseases manifest themselves."[40] And a 2007 review article about gender difference in posttraumatic stress disorder says the following: "Women's greater biological complexity compared with men's implies substantial gender-related differences in HPA-axis regulation. The HPA axis and the female reproductive system are complexly intertwined. Because females have greater variations in neuroendocrine responses because of their reproductive cycle, their data often present confusing patterns of results."[41] It is worth noting in these recent studies—both of which address the question of why women suffer higher rates of mental-health diagnoses than men do—that the female menstrual cycle is depicted as something that has introduced complexity, as if the menstrual cycle itself is an outside force that has been imposed on an otherwise normal (i.e., male) body. As indicated in the above quotation from the 2013 review article, evolutionary narratives are often used to account for this purportedly greater complexity of the female brain, which is thought to be a result of the female brain's ongoing need to adapt to the fluctuation and change that are reportedly introduced by the menstrual cycle.

Evolutionary narratives like these assume many different forms in other lines of recent research, sometimes leading to arguments that, for the first time in recorded history, emphasize some positive effects of hormones on the female brain. For example, a 2008 review article titled "Reproduction-Induced Neuroplasticity: Natural Behavioural and Neuronal Alterations Associated with the Production and Care of Offspring" refers to the "multiple exertions that define motherhood." As Craig H. Kinsley and Kelly G. Lambert say in this article, "a new mother is faced with many novel and unforgiving challenges, and any improvements in the behaviours on which she herself relies for sustenance would be advantageous for her, for her genetic offspring, and for her genetic legacy." The authors go on to describe how the female brain in particular has continually been modified throughout the life cycle, using more positive language than we often see in scientific texts on this topic: "In particular, the adult female brain is significantly modified, and reorganized in some examples, in surprisingly rapid fashion, as during the oestrus cycle and after exposure to gonadal steroids." The authors briefly summarize the different topics that have been studied regarding "endocrine-neurone relations that may be unique to pregnancy," and they state that the purpose of their review article is to observe "the dynamic female brain and its natural changes that accompany reproduction and that regulate the female's pup-directed, social and ancillary behaviours." In describing the motivation for their interest in this topic, the authors refer to rats and human mothers side by side: "Closely observing the basic interactions between a mother and her offspring, rat as well as human, inspired us to ask the simple question about the onset of motherhood and its requirements for the new mother, including a facility for rapid learning of spatial environments for the maternal female." They summarize their review findings as follows: "Our data suggest that the hormones of pregnancy interact with postpartum pup exposure to produce an 'enriched environment' and thereby improve learning ability (at least temporarily and possibly permanently), mitigate anxiety and stress responsiveness, and enhance problem solving in a novel context (integration of cognitive and stress responses) in the mother."[42]

Kinsley and Lambert's review article is notable for several reasons. First, it exemplifies a strand of recent research that emphasizes more positive effects of hormones on the female brain. Second, the authors acknowledge a complex of factors—including hormones and the experience of interacting with the pups— as possible forms of influence leading to the increased neuroplasticity of the brain that they conclude occurs with parenting (to a greater extent in females and, to some extent, in males). They summarize these findings as follows in the conclusion, "in neuroanatomical realms, we have observed many indices of plasticity that suggest a brain in fuller expression of its abilities." The article closes with a broad-sweeping claim that "in total, our research, when taken together with that of others, has elucidated a portion of the many changes that naturally occur in the transition from nulliparity to motherhood, potentially providing valuable information relating to the development of normal and abnormal maternal interactions, as well as the underlying mechanisms contributing to perinatal and subsequent mood disorders."[43] For all of these reasons, Kinsley and Lambert's review article can be seen to suggest the possibility of eventually escaping some of the long-standing assumptions that we have seen in previous research about the problems that hormones caused for the female body and brain.

However, to understand fully how recalcitrant are these long-standing assumptions and to see the effects of enthymeme in this body of research, we must consider how Kinsley and Lambert's review article has been cited in the subsequent literature. For instance, Laura M. Glynn's 2012 article titled "Increasing Parity Is Associated with Cumulative Effects on Memory" cites Kinsley and Lambert's review article, among other texts, to support the claim that "animal models have demonstrated that the effects of reproduction and mothering on cognitive function and brain structure are additive and persist throughout the life span."[44] The study that Glynn's article summarizes tested 254 women for verbal recall memory performance four times while pregnant and at three months postpartum. Participants' memory was tested with a "paired-associates learning task," which entailed presenting each woman with twelve pairs of unrelated words and then, immediately afterward, scoring the woman's ability to recall the second word in each pair after being shown only the first word. The results showed that participants' capacity for verbal recall diminished during pregnancy and that the effect was cumulative, so that the memory performance of mothers with greater numbers of children was more greatly impacted. Rather than presenting this as a completely negative effect on cognitive performance, however, Glynn draws on previous studies of rodent and human mothers—including Kinsley and Lambert's review, among others—to suggest that this diminishment of verbal recall memory might be a trade-off in which the cognitive detriments of motherhood are outweighed by some cognitive improvements: "There is some evidence from the rodent model that suggests memory decrements may represent the cost of the architectural remodeling of the maternal brain."[45] In the conclusion, in which the author discusses the research implications, we see these ideas most fully elaborated:

> In the realm of cognition, a potentially fruitful pathway would be to investigate cognitive functions more directly relevant to caring for a new infant. Taking a broader evolutionary perspective, verbal recall memory, which is compromised during and after pregnancy, is a relatively recent adaptation and while essential for performing in the modern workplace, for example, it is likely to be less central to success in caring for an infant. However, such skills as attention to infant cues, the ability to multitask (or divide attention), and the ability to detect threat might be enhanced by pregnancy and child rearing.[46]

In this paragraph, we see some old ideas resurface in a manner that is relevant to this chapter's study of enthymeme. This text is enthymematic insofar as it occludes any explicit discussion of a whole set of value judgments about which cognitive skills are "essential for performing in the modern workplace." By taking those value judgments for granted, this argument draws us back into long-held assumptions about the kinds of activities that female brains are most ideally suited for. The author cites no sources to support this distinction between different types of skills, but instead expects us to take for granted that successful people in today's workplace do not need to be adept at "such skills as attention to infant cues, the ability to multitask (or divide attention), and the ability to detect threat." Of course, "attention to infant cues" per se may not be all that valuable in a workplace (other than childcare facilities), but there is no reason to dismiss multitasking and detecting threat as skill sets that women (or men, for that matter) do not need in today's workplace. And even "attention to infant cues" could be viewed differently if it were phrased more along the lines of "responding to the needs of employees and colleagues" or "displaying empathy." The only way that this argument works is by accepting a set of unstated assumptions about the kinds of cognitive skills that women need in order to be good women.

To appreciate fully how this enthymeme plays out, we must consider recent feminist critiques that expose additional dimensions of the mixed blessings that have arisen from recent research that suggests a more positive interpretation of the cognitive effects of pregnancy hormones. For example, Davi Thornton uses the term *mommy economicus* as the basis for a feminist critique of the new discourses of neuroscience in recent years that have tried to reclaim motherhood as having a positive influence on the brain. Thornton observes that although these new discourses are potentially empowering to mothers, they also place new burdens on today's mothers by suggesting that to maximize the cognitive benefits of pregnancy and mothering, mothers would have to take certain actions or behave in particular ways. As Thornton says, "these discourses construct the ideal mother as a privileged exemplar of neuroplasticity, or the idea that the brain (and hence the self) is malleable and can be enhanced through work on the self." Thornton emphasizes trends such as the increasingly blurred boundaries between home and work, and the idea that today's technologies transform "all spaces into potential workplaces."[47] As Thornton summarizes, "because of the extraordinary plasticity of her brain, mommy economicus is agile, flexible, and adept at directly leveraging the customary practices of motherhood—including caring for both her home and her children—for corporate profit and personal development."[48] In short, even though the cognitive damage supposedly inflicted by pregnancy hormones has always been portrayed as something over which mothers had no control (because it was caused by hormones), the cognitive benefits of pregnancy hormones are portrayed as something that could materialize only if mothers engaged in specific practices that made her an ideal neoliberal subject.

Thornton refers to "popular neuroscience" as a set of discourses that perpetuate the idea that the self is the brain, and that circulate complex scientific findings related to neuroscience in ways that can be taken up and incorporated into popular understandings of health and wellness. In these discourses, neuroplasticity is emphasized as an ideal that we all should strive for, and the maternal brain is increasingly seen as exemplifying this trait of neuroplasticity. Thus, Thornton defines the "mommy brain" as "a special organ endowed with supreme powers of plasticity."[49] In the mommy-brain literature, Thornton says, hormones play an important role: "Often a primary culprit in medical accounts of female pathology, the mommy brain story casts hormones as beneficent protagonists who prepare women's brains for enhancement by triggering neuroplasticity." The maximum benefits of neuroplasticity are not guaranteed, however. As Thornton says, "hormones set the groundwork for the construction of the mommy brain, but they must be accompanied by a second input—positive early motherhood experiences."[50]

Like so many discourses related to gender difference, Thornton is correct to point out that these new discourses about mommy brain are a double-edged sword. Thornton's observations relate in an interesting way to the enthymematic pattern in Glynn's study that I noted above: an enthymeme that is perpetuated by the omission of any explicit discussion of the values that are used to distinguish between skills that are needed to succeed in the workplace and skills that are needed to care for an infant. These new discourses that Thornton analyzes in many ways resonate with old, essentialist ideas about motherhood as the quintessential means by which women realize their potential as humans. But when we look more closely, it also becomes clear that these new cognitive traits that are idealized are not strictly biological traits that new mothers passively inherited; rather, these traits must be maximized and enhanced through specific actions that mothers would take and the ways in which they would approach motherhood. Perhaps not surprisingly, these actions are ones that were likely available to mothers of a certain income level and socioeconomic group because they involved such actions as regular exercise, leisure activities away from the baby, and so on. An important part of this discourse is that, perhaps for the first time in history, the same capacities and behaviors that were idealized in a good mother were seen as serving mothers in the workplace. As Thornton says, "motherhood becomes an entrepreneurial practice (or set of practices) that is continuous with, and in many cases identical to, the entrepreneurial practices of work." Thornton discusses some of the specific gendered traits that have long been seen as troublesome in the discourses about women's hormones, such as women's "ever-changing reproductive systems [that] make them fickle, flighty, and unstable," but she notes how the new mommy brain discourses turn these around so that "malleability is an asset rather than a curse, and accusations of fickleness give way to celebrations of flexibility."[51] As in the Glynn study, the enthymeme at play in these discourses is built around an occlusion of any of the social, cultural, and economic factors that inform our value judgments about which kinds of cognitive skills are valued in which situations.

Reflecting the concerns that Thornton's feminist critique raises about these new discourses that celebrate the maternal brain, many of the scientific articles that I have examined in this chapter perpetuate this emphasis on the need for the surveillance of the maternal body and brain. When the authors address the implications of their studies, for instance, there is often a suggestion that pregnant women should be warned of the likelihood that memory loss will occur, and that they should be given strategies to prevent that memory loss from interfering with their daily lives. For example, despite the uncertainty that they acknowledge in the available scientific evidence, and despite their acknowledgement of a need for further research, Brett and Baxendale use bold and explicit language in making clinical recommendations based on the body of evidence that is summarized in their review article. Specifically, these authors suggest that "a large proportion of mothers may be at risk of GMI and antenatal care may be improved if women are formally prepared for GMI prior to conception or in the early stages of pregnancy when they can be helped to develop memory strategies to overcome the impending difficulties."[52] Henry and Rendell's review article also includes similar language about the clinical applications of evidence presented in their meta-analysis: "These data should help to guide the interpretation of neuropsychological data for the purpose of determining cognitive status in individuals who are pregnant or postpartum."[53]

This emphasis on the need for physicians, and women themselves, to surveil the pregnant mind and body is, of course, concerning: If women need to be warned of the risk of cognitive change even before they become pregnant, does this not suggest that pregnancy itself comes with a risk that women will be less competent in their professional duties? Is it supposed to be reassuring that they can work with their physicians or other health-care professionals to develop strategies to help compensate for this deficiency? And when the science suggests that the positive effects of pregnancy hormones on the maternal brain can be maximized through specific actions that mothers themselves take, does this place an additional burden on women who are already working hard to negotiate conflicting demands in a society that does not have many structures in place to support mothers? In light of questions such as these, we must proceed with caution in evaluating the newest discourses of the hormonal woman, even when those discourses proclaim the positive rather than negative effects of female hormones.

Conclusion

This chapter's analysis continues the book's argument that the existence of hysteria—which persisted in so many different forms for so many centuries—created a space in scientific discourse that hormones easily filled. If we think of enthymeme as functioning like the eye of an argumentative storm, we might say that the enthymematic component of these long-standing arguments has always been present in that quiet, unchanging space, but something new was needed to fill that space when hysteria no longer worked as an explanation for women's mysterious behaviors; filling the space is precisely what hormones have done. To say that a component of these long-standing arguments is enthymematic is to say that this component has been omitted for a specific reason. That is, because audiences do not need any evidence to be persuaded about a given point, the evidence has been left out, which enables these arguments to be simpler and to resonate with the various audiences who need no convincing about the shared assumptions that support the omitted piece of the argument. The current rhetorical theory of the enthymeme, as it has been employed in traditional rhetorical scholarship, demonstrates that this purposeful omission makes the argument more persuasive than it would be otherwise. But, as we have seen in previous chapters, the case of female hormones also causes trouble for traditional rhetorical concepts that have been applied mostly to subject matter that is far removed from the messy, unruly realities of the female body. Extending beyond this traditional theory of the enthymeme, then, we see in the case of female hormones that it is not just any piece of the argument that is omitted, but it is specifically a piece of the argument that, if it were to be filled in, would cause the whole argument to crumble. Jackson notes a similar situation in the context of race, arguing specifically that enthymemes can function to uphold white supremacy:

"Racist enthymemes can function to support white supremacy inconspicuously and indirectly,"[54] allowing value judgments that support white supremacy to remain hidden behind the scenes of arguments that would otherwise seem logically sound.

Likewise, in the three lines of research examined in this chapter—research about pregnancy brain, research about women's mental health, and research about the cognitive benefits of female hormones—the missing piece that constitutes an enthymeme functions to uphold the tendency of Western science to privilege male over female biology. If these missing pieces of the arguments were filled in, these arguments would simply not work. So we expand our theoretical grasp of the enthymeme by highlighting cases in which the omission that constitutes enthymeme in an argument is something that not only makes the argument more compelling but is also crucial for the argument to exist as such. Without the empty space that these enthymematic omissions allow to exist, these arguments would not have the momentum—the capacity to move audiences both close to and far away from them—that they have been proven to have.

In this regard, enthymeme as employed in this chapter offers yet another way to visualize Serres's topological movement in the context of scientific progress. Thinking in terms of topology entails acknowledging the spatial dimension of scientific progress, in addition to the temporal dimension, and as such, we might envision the eye of the argumentative storm that enthymeme creates as a space that remains constant in these discourses, whether it is filled by a wandering womb, a hurrying pig, chemical messengers, or hormones. Based on my analyses of the most recent hormonal discourses that have been presented in this and the previous chapter, yet another layer to this problem is exposed—namely, that because hormones slipped in and took the place of hysteria, a long history of argumentative momentum was already built up to support the idea that something inside women caused them to behave less rationally and predictably than men behave, as well as the related ideas that women's bodies have created interference for their minds in a way that was never suspected to be the case for men. These analyses of the most recent discourses help us understand exactly how the discovery of hormones—and the incorporation of hormones into the discourses used to explain these phenomena—has transformed the science and popular understandings that surround this set of behaviors and symptoms that have long been characterized vaguely as "female problems." Continuing one of the larger arguments of this book, such analyses can also help us see how this discovery initially had the potential to transform these understandings in a way that would empower women, and some of the reasons why this potential was never realized.

The common theme across the numerous strands of scientific inquiry addressed in this chapter, and the ways in which their findings are reported in the media, is that even if we accept the conclusion that pregnancy or some other hormonal event in women's lives induces some diminishment of memory or some other form of cognitive change, the studies are usually not designed in a way that would lead to the scientific discovery of the precise mechanisms through which hormones would have caused these problems. However, scientists and the journalists who report their findings to the public are both quick to assign hormones as the cause in almost every case, regardless of the particular hormonal event being studied or the particular form of cognitive impairment in question. By exposing some of the ways in which hormones have come to function as an enthymematic element in scientific arguments about women's bodies and brains, we can see how hormones have come to fill up the argumentative space when no other logical explanation can be found. In short, because of public and expert audiences' preexisting faith in the power of hormones, hormones have come—without too much additional explanation or support—to be accepted as a plausible cause for many of these effects. Furthermore, this chapter's analysis has emphasized that hormones can function in a similar way, regardless of whether their purported cognitive effects are depicted as negative or positive.

It is interesting to note that the term *enthymeme* has something of a kinship with *hormone*. As rhetorical scholars have observed, *enthymeme*'s root word *thymos* translates as "'heart,' meaning the seat of emotions and desires, or of motive, of the sometimes uncontrollable forces of desire and wish that drive human intentionality" and "is, moreover, often linked to both the production and the reception of passional thought and eloquent, persuasive discourse."[55] Thus, both concepts refer to entities that are understood to incite movement that is outside the realm of rational control. If *hormone* is that which sets in motion those bodily processes that are not subject to control by the brain and central nervous system, *enthymeme* is that which sets in motion elements of persuasion that cannot be understood exclusively through logic or rationality.

In this regard, the notion of enthymeme also relates to Serres's observations about the herd mentality of academic knowledge production. Although much of the emphasis in recent thinking about such knowledge production has been about the social construction of knowledge, Serres suggests that there might be value in encouraging a more individual approach: "New ideas come from the desert, from hermits, from solitary beings, from those who live in retreat and are not plunged into the sound and fury of repetitive discussion. The latter always makes too much noise to enable one to think easily. All the money that is scandalously wasted nowadays on colloquia should be spent on building retreat houses, with vows of reserve and silence. We have more than enough debates; what we need are some taciturn people."[56] As unrealistic as Serres's suggestions might be, his remarks offer an appropriate closing place for an analysis that has exposed the damage that can be done by too much "groupthink" in science and the difficulties that are encountered in overcoming centuries of argumentative momentum about a topic as politically fraught as the effects of hormones on the female brain.

8

From Hysteria to Hormones

For at least a couple of decades before Starling first used the word *hormone* in a 1905 lecture to the Royal Society in London, experts knew that a chemical substance enabled the organs to communicate with each other to enable processes like digestion and respiration. However, they did not have a good term to describe these substances; they used vague words like *chemical messenger* and *internal secretion*. None of these words was powerful enough to win the argument, so the experts went back and forth, quibbling over how to interpret the same old evidence. The term *hormone* was a game-changer. The chemical substances that the new word named had been known to scientists for a long time. But when they finally had a word that they could agree on, the science moved forward after decades of standstill. By 1915, endocrinology had become established, and this field continued to experience rapid growth for many years after that.

Thus, one of the stories told in this book is about the founding of an entirely new medical field—endocrinology—and of a physician who became famous because of the rhetorical movement that he was able to effect after those decades of standstill. The science that Starling reported in his 1905 lecture was not very new at all, but because he used the term *hormone* for the first time, he changed the world. And he is remembered in the history of medicine as the person who discovered hormones, even though many scientists before him labored for years in their laboratories to understand the internal mechanisms of communication that enabled bodily functions such as respiration, digestion, and reproduction.

As compelling as this story of Starling's rhetorical accomplishment might be, it is only one small part of the larger narrative that has been told in this book: the centuries-long narrative of a shift from the hysterical woman to the hormonal woman as the primary metaphor that we used to explain almost everything that could go wrong with women's physical or mental health. The belief in hysteria, which has a history that spans the centuries, was based on wild imaginings about the behavior of the womb deep inside a woman's body, whereas the relatively new belief in hormones is based on scientifically verified chemical substances with resulting behaviors and systemic effects that can be measured, documented, and replicated in the laboratory. The short version of this story is that science has gradually come to replace mysticism and religious beliefs as the basis for understanding women's bodies and women's health.

As I have argued, however, the transition from a hysterical-woman metaphor to a hormonal-woman metaphor has been far less absolute than previous approaches to the history of medicine might lead us to expect. This continual process, in which new terms and concepts have gradually morphed from older terms and concepts, offers us new ways to comprehend the rhetoric of science as a form of movement that is characterized by anything but a progression along a straight timeline. The forms of movement that are evident in the scientific rhetorics that I have analyzed in this book are best characterized as folding, fluxing, morphing, and twisting. Thus, through close examination of the scientific and popular texts that facilitate these forms of movement, we can see how a concept such as hormones never really breaks from its history but, instead, comes to encapsulate key ideas from that history, reshaping these concepts in ways that fit the demands of ever-changing rhetorical contexts. This highlights a fundamentally conservative element of the scientific endeavor, suggesting that one of the reasons why new ideas emerge is to preserve old ways of thinking— to make those old ideas acceptable to new audiences—not only to effect a clean break from the past.

Perhaps it is not coincidental that much of the energy of medical experts during the many centuries that this book covers has been devoted to figuring out how and why the womb refused to stay put inside women's bodies. It is almost as if this entity inside women that has been depicted as unpredictable and hard to tame—sometimes even referred to in the ancient texts as a wild animal—mimicked the kind of topological movement of scientific thought that I have tried to articulate as this book's contribution to the rhetorical history of the female body. Just as the womb's refusal to stay put inside women's bodies— and its unpredictable patterns of movement—caused numerous problems for the scientific and medical experts who have tried for many centuries to figure out the female body, this book's analysis has tried to cause problems for traditional rhetorical understandings of concepts such as discovery, invention, memory, metaphor, enthymeme, and stasis. Each of these concepts has been employed in this book's analysis as a way to visualize the precise form of movement that appears to be most salient in the rhetorical activity that occurred at each moment that I have highlighted in this long history. To weave together the threads of insight that these chapters offer into the movement that was occurring at particular moments in this long rhetorical history, I would now like to turn to the rhetorical concept of kairos. If we understand the story of scientific change told in this book as a matter of kairos—an approach that many rhetorical scholars have taken in analyzing the history of science—we might emphasize that when a new term such as *hormone* emerges, it takes hold only if the time is right for that term to resonate with its expert and popular audiences. Or, if not, then the term does not take hold right away, and it might instead take hold later, when the time becomes appropriate. According to this line of thinking, terms, concepts, or arguments resonate with audiences at specific times and places, and if they do not resonate, then they die out—possibly forever, possibly for only a short time, or until there is another moment when the time is more appropriate. This analytical approach has offered useful insights into the forms of rhetorical movement that characterize key moments in basic science, such as the discovery of DNA,[1] and into the development and change that occurs over centuries in regard to treatments of specific medical conditions.[2]

In many ways, this book's topological approach to the history of science and medicine resonates with rhetorical theories of kairos as these rhetorical theories have been employed in previous rhetoric of science scholarship. Rhetorical critics have invoked kairos to illuminate scientific changes to a number of scientific and medical topics,[3] and this concept has proven useful perhaps because, as Miller observes, kairos "compresses into one handy Greek term a variety of related but distinguishable rhetorical considerations."[4] When we take kairos as our guiding concept, rhetorical critics emphasize the matter of timing and how that determines the success of a given argument. Thus, for instance, Miller compares Oswald Avery and his collaborators' 1944 report about the role of DNA in the genetic transformation of bacteria to James D. Watson and Francis H. C. Crick's 1953 *Nature* paper that reported the double-helix structure of DNA nine years after Avery and his collaborators' article. Although Avery is given credit in retrospect for being the first scientist to identify DNA as the key genetic substance, Avery's argument did not receive nearly as much recognition by his contemporaries as did Watson and Crick's article nine years later. As Miller observes, the dramatically different responses to these two different articles can be understood in terms of timing. Avery's cautious style might be seen as a necessary adaptation to the kairos of his time, whereas Watson and Crick's bolder style was appropriate for their time. This is a powerful counterargument to those who might want to offer the reverse interpretation and say that Watson and Crick's argument was more successful *because* of their bolder style.[5]

In short, the sophistic concept of kairos has been enacted in contemporary rhetoric in a way that depicts time as something that evolves and changes rather independently of the rhetor's moves. If the rhetor presents an argument in the right way at the right time, it will be an effective argument. There is much guess-work and much variation, but when the right argument is presented in the right way to the right audience at the right time, that argument can be judged as effective. The successful argument is akin to the archer's arrow that hits its target, and in retrospect, as a rhetorical critic, it is possible to look back and make some combination of claims about a rhetor's success in hitting a target. Depending on which theory of rhetorical exigency and audience the rhetorical critic espouses, rhetorical success might mean that the rhetor created the target in a way that would be easiest for the "arrow" to hit, or it might mean that the rhetor accurately perceived the precise location of that target and then crafted an argument that would be certain to hit it.

In many ways, this formulation of kairos as it has been employed by Miller and other rhetorical scholars resonates with Serres's conception of time as "turbulent and chaotic."[6] In fact, as Miller observes, the concept of kairos has its origins in sophistic rhetoric, particularly in an emphasis on the confusion, chaos, and indeterminacy that are inherent in any rhetorical situation. In this way of thinking about rhetorical situation, there has to be some confusion and uncertainty in order for someone to perceive that rhetorical action is necessary.[7] In contrast to Miller and other rhetorical scholars who have theorized kairos, however, Serres asks us to adopt a more dynamic, less predictable, understanding of scientific changes as they unfold over time. He offers a few metaphors that sharpen our awareness of these dimensions of scientific change. As Serres says, "time flows like the Seine, if one observes it well. All the water that passes beneath the Mirabeau Bridge will not necessarily flow out into the English Channel; many little trickles turn back toward Charenton or upstream."[8] From this metaphor of the River Seine, Serres derives the metaphor of "percolation," a word that, as I have discussed in chapter 5, he says is more accurate than verbs such as "flowing" or "passing" that we often use to characterize the progression of time. As I have demonstrated, Serres encourages us to see the continuity and stillness that reside within big moments of rhetorical movement in science. This fluid and multidimensional approach is especially appropriate in understanding the forms of rhetorical movement that are involved in a concept such as that of the hormonal woman, which did not emerge at a specific moment, but evolved from the earlier concept of the hysterical woman in a process that entailed multiple moments of scientific movement and occurred over a span of several decades.

In light of the events that are narrated in this book, then, we must expand our thinking about kairos in ways that enable us to think of the right moment, or the right time and place for an argument, as something that is multiple and unfolds over time—similar to Serres's notion of percolation, or his metaphor of the River Seine with its deceptively smooth surface. Rather than one right moment for a particular argument, we might conceptualize a series of moments that are not connected in any predictable or rational way. Any of these moments might be the right time for a new argument to take hold, and in many cases, that seemingly new argument might, in fact, be the result of very old ideas as they percolate down through time and find a new expression that works for the present moment.

Additional theoretical frameworks that rhetorical scholars have used to illuminate the history of science include Thomas Kuhn's concept of paradigm shifts, and the related concept of incommensurability, that were theorized both by Kuhn and by Paul Feyerabend.[9] These concepts offer different frameworks for understanding the moments when scientific change occurs, and they have received a great deal of attention from rhetorical scholars who have both extended and reshaped our understandings of these concepts in relation to the rhetorical history of scientific subjects. Some of this scholarship offers insights that intersect with this book's insights into the manner in which new ideas can incorporate elements of the old. For example, Campbell's in-depth studies of Darwin illuminate how Darwin's ideas emerged from those of his predecessors, even though he made a radical break from them, and in so doing, Campbell causes us to question whether incommensurability is as real as we might think, even in the scientific controversies in which disagreement appears to be most profound.[10] But Campbell ultimately does not question the idea of scientific progress. He still concludes that Darwin "won" the evolution argument,[11] beating out his predecessors in a manner that suggests an underlying assumption of advancement and linear progress that is seriously questioned by Serres's notion of time as topological. Even as they try to rethink concepts from history and philosophy of science such as paradigm shifts and incommensurability, rhetorical scholarship is still largely framed in the assumption that science advances through a series of breaks or ruptures that allow new ideas to be detached from those that came before. This book's topological analysis, by contrast, highlights a fundamentally conservative element of science that is easily overlooked in a world where science is idealized as the mode through which knowledge perpetually improves.

When we think in terms of topology, we emphasize what lies beyond the smooth surface of those rhetorical situations that constitute the moment when scientific change appears to occur. So when a term like *hormone* arrives at such a perfect time—made perfect because the term *hysteria* was losing its diagnostic relevance and needed to be replaced—we might look beyond the surface and see that this new term was actually long overdue. Decades earlier would have been a more ideal time for the term *hormone* to replace vague terms such as *chemical messengers*. If the term *hormone* had emerged earlier, the more realistic understanding of the body's internal communication processes that became possible with the ensuing establishment of endocrinology could have saved thousands of women from being hospitalized and subjected to bizarre treatments for a condition—hysteria—that was never actually proven to exist. Viewed from this perspective, we must dramatically expand our thinking about concepts such as kairos to account for the acknowledgement that there might not be a single moment that is inherently a good or bad time for a new scientific term or concept to emerge. Furthermore, what seems like the perfect time for a new term or concept to take hold might not actually exist as a single point in time, like an arrow hitting its mark. It is, rather, a confluence of the different arrows that previously tried to hit the mark but missed. And, in fact, closer scrutiny reveals that there is not a single mark, but rather a shifting target that changes as the people, places, and things around it change. When we consider all these imperfections and harsh realities that fold into a moment that seems like "the right time" for a concept such as hormones to emerge, instead of focusing just on the smooth surface of that single moment, we can appreciate more fully how a new concept such as that of hormones contains within it bits and pieces, or fragments, of the older terms or concepts that it replaces. This approach might ultimately help us to respond to the challenge that is posed by Ballif, who asks rhetorical scholars to rethink our existing approaches to historiography so that we are not always focusing on a series of distinct events that are presumed to follow each other in a linear fashion that can be retrospectively traced by the rhetorical critic. As Ballif asks, "how might one write a history that attends to moments of rupture—without, and this is the key—pointing to how these moments of rupture instantiated a new series of moments or ushered in a new rhetorical moment?"[12]

The persistence of ancient ideas about the wandering womb in medical and popular texts that were written during the nineteenth and early twentieth centuries offers a compelling response to Ballif's question. It also gives life to Serres's assertion that time is topological—that is, perpetually folding in on itself, rather than moving endlessly forward in a linear direction. Regardless of the differences that I have traced throughout the historical eras, an important element that has stayed the same, even in today's science, is the notion that there is some direct link between women's reproductive organs and their brains. As we have seen, today that is a much more finely tuned and scientifically supported notion because, thanks to Starling's 1905 lecture, we have the concept of hormones as chemical messengers to link the two. But well before the emergence of the term *hormone*, we had centuries of textual evidence to document the persistence of this belief—twisted, reshaped, and contorted as necessary to fit each historical time and place—in a direct connection between the female reproductive organs and the female brain.

Thus, if we think in terms of topology, we look at what stays the same even while these larger changes occur, which does not mean doubting these larger changes. Those changes are easy to document. Furthermore, as I mentioned above, obvious improvements occur as a result of scientific advancement—most notably, that "hormonal" women are not being locked up in the Salpêtrière in the way that "hysterical" women were in the nineteenth century. What I have aimed to accomplish by rereading rhetorical history through the lens of topology is perhaps most accurately characterized as a shift in perspective, so that we are able to see the minutiae that remain constant even against the backdrop of apparently revolutionary changes.

Another important consideration is that, through the lens of topology, we are afforded a broader view of historical change as it extends through multiple eras, even extending back to ancient texts, and this broader view reveals that the experts have been making claims about scientific progress for a long time; the situation is not unique to the late nineteenth and early twentieth centuries, which are often held up as the eras in which experts started to demand tighter affiliations between science and medicine. Thus, topology offers a complementary (not oppositional) view of scientific and medical change in relationship to kairology and some of the other theoretical frameworks that rhetorical scholars have used to conceptualize change in science and medicine. Taken together, these concepts might enhance our ability as rhetorical scholars to make increasingly well-defined contributions to ongoing scholarly investigation in fields such as philosophy and the history of science and medicine.

Along these lines, I hope that this book's rhetorical analysis also sheds new light on the disappearance of hysteria, which in itself has been a subject surrounded by much darkness and mystery, even in the extensive scholarly attention that it has received from historians of medicine. In fact, previous historical research offers some speculations about this disappearance, but no conclusive answers. For instance, in her cultural history, Veith notes, "it is often stated by psychiatrists and other physicians that the disease has ceased to exist."[13] Given that Veith's text was published in 1965, she was writing at a time when hysteria was still widely acknowledged and addressed in psychiatric textbooks, although she notes that the term *hysteria* was removed in 1952 from the *Mental Disorders Diagnostic Manual* of the American Psychiatric Association.

More recent historians fill in details where Veith's analysis left off. For instance, Garry Kinnane notes that in the early twentieth century, hysteria began morphing into an array of mental disorders, including anorexia nervosa and bulimia, among others.[14] Offering a slightly different perspective, Micale identifies two competing explanations that historians have offered for hysteria's disappearance. The first explanation revolves around "psychological and sociocultural factors." In this view, "the disappearance of hysteria is the result of de-Victorianization."[15] In other words, the symptoms disappeared because the sociocultural factors that caused them disappeared. The second explanation is the "psychological literacy" explanation, which suggests that hysteria disappeared when people gained a better vocabulary for describing their problems; with this better vocabulary, there was less need for the somatization of anxieties that had frequently led to hysteria in earlier eras. Micale says that both of these "sociogenic interpretations" are inadequate, and that we "need to look elsewhere for answers." He claims to offer a different understanding—namely, that what looked like hysteria's disappearance was actually a process of reclassification—and goes on to say that the important causes for this disappearance were "scientific factors involving biomedical discoveries in etiological theory and diagnostic technique" that "were reinforced and accelerated by a series of historically specific sociological factors."[16] Echoing some of my conclusions in earlier chapters of this book, Micale then notes that despite the progress that was made during the nineteenth century in discovering etiologies for many other diseases—such a discovery did not happen in the case of hysteria—so physicians kept explaining this disease in terms of its symptoms rather than its identifying causes. Micale's focus, which is similar to

Kinnane's, is the wide array of mental disorders that came to replace hysteria. He rightly points out that with the growth of new disciplines such as psychology, physicians gained access to a much more robust vocabulary for classifying the wide array of mental illnesses. Thus, hysteria was replaced by terms that indicated "specialized categories" such as "hysterical mania" and "hysterical insanity."[17] Specific diagnoses that came to "incorporat[e] a number of clinical constituents of the former hysteria" included schizophrenia, and another trend that he identifies is a redefinition of hysteria in the early twentieth century in what he describes as "a retention of the word *hysteria* but its application in a new diagnostic context." As he says, hysteria came to describe "a transient psychological reaction" or "a pattern of symptom formation that may appear in conjunction with other more basic psychopathologies."[18] This historical insight likely explains some of the uses of the terms *hysteria* and *hysterical* that we see in the articles that I have analyzed in chapter 6's tracing of the gradual shift in metaphors that resulted in a mixing of the hysterical- and hormonal-woman metaphors in medical texts of the early twentieth century.

Although there are certainly some similarities between Micale's conclusions and the arguments that I have presented in this book, I still contend that a rhetorical analysis can offer something more. Rather than just note the existence of the word *hysteria*, and variations on it, that appear in mid-century medical texts, a rhetorical analysis such as that which is presented in this book can carefully identify the intricate processes through which specific uses of language contributed to, and interacted with, larger processes of social and scientific change. Furthermore, in a rhetorical analysis, we start from the position that these social, rhetorical, and scientific dimensions of change are closely connected with, not separate from, each other. By contrast, in historical analyses such as those that are offered by Kinnane and Micale, there is no evidence of a move beyond specific diagnostic categories to account more fully for the extent that the hysterical woman served as a metaphor for women's health problems in a broader sense.

In addition to these contributions to the history of medicine, this book's rhetorical history relates to other important themes that feminist historians and philosophers of science and medicine have exposed in studies of similar subjects. For instance, it is agreed among these feminist historians and philosophers of medicine that from the earliest days of sex hormone research, female hormones received far greater attention than did male hormones. This disparity is reinforced in my analysis and, in fact, I have demonstrated in chapter 5 that this disparity can even be documented in the text of Starling's 1905 lecture. How to interpret this disparity between the amount of scientific attention that has been devoted to male and female hormones, however, has been a subject of debate between the two main feminist scholars of the history of sex hormones, Oudshoorn and Roberts. Both of these scholars document that much greater scientific attention was given to female hormones than to male hormones in the early twentieth century, but they offer different interpretations of this fact. Oudshoorn offers a technoscientific explanation, emphasizing the purely practical reasons why chemical versions of female hormones were synthesized earlier and more easily than was the case for male hormones. For a variety of reasons, according to Oudshoorn, chemical substitutes for female hormones became available much sooner and more easily than was the case for male hormones.[19] Roberts contends, however, that this matter of availability does not sufficiently answer the question. Offering a political explanation to counter Oudshoorn's technoscientific explanation, Roberts insists that we also must consider the politics surrounding the emergence of gynecology as a new subfield in the early twentieth century, and we must ask, "why were clinics set up around women's bodies and not around men's? Which *practices* of masculine supremacy, or . . . other systems of structured inequality' were built into this practical fact?"[20] Roberts argues, along lines that are similar to my analysis in chapter 6, that misogynistic ideas about women's bodies as inferior to men's bodies became deeply entrenched in the 1930s, when it first became possible to use blood tests to detect the variations that occur in women's normal hormone levels, even throughout normal menstrual cycles. This makes women, in contrast to men, appear to be unstable and unreliable, Roberts contends. The efforts of gynecologists and physiologists alike to medicalize women's bodies, according to Roberts, were closely tied to the new possibility of the use of chemical treatments on women's bodies—hormone-based drugs that then became cornerstones of gynecology: "Sexual difference was now seen to be located in chemicals rather than in organs or cells."[21]

Regardless of which interpretation we accept, the science of hormones played an important role in this early twentieth-century quest for the control of women's bodies. Women's own hormones, from the earliest days of the discovery of hormones, were perceived as problematic because they got in the way of medical control over women's bodies. Only a few decades after Starling coined the word *hormone*, however, artificial hormones started to offer the promise that medicine could take back that control. More recently, psychologist Pedro Pinto has echoed these ideas, observing that in expert and popular understandings of adolescent sexuality, hormones have taken on dramatically different meanings for boys and girls. Whereas boys' hormonal influences are depicted quite simplistically as centering on testosterone, girls' hormonal influences are depicted as much more complex, so that girls are instructed to pay attention to the different hormones that impact them at different phases in their menstrual cycles. Thus for girls, Pinto observes, "knowing one's own hormonal truth becomes an imperative care in the art of self-government, through which the 'natural' costs of femininity may be minimized."[22]

Although the approaches of these previous researchers would seem to beg for an investigation of the rhetorics surrounding these topics, none of their analyses focuses explicitly on rhetoric. As I have demonstrated in this book, there is much space for rhetorical analysis to build from these previous historical and philosophical analyses. A rhetorical approach is ideally suited to advance the conversation in this manner because, as rhetorical scholars, we are attuned to the nuances of language that experts in various fields, as well as nonexperts, use to talk and think about aspects of the body, whether in scientific texts, popular texts, or other forums like today's online venues where people go to talk about health-related topics. A rhetorical approach to medical topics such as hysteria and hormones adds to previous historical scholarship by examining specifically how various groups and individuals have persuaded, and been persuaded, by arguments that have emerged and evolved throughout the centuries. As a result, in this book I have shifted focus away from the either-or approach that has unfolded in the debate between Oudshoorn's technoscientific and Roberts's political explanations of sexist bias in hormone science. Rhetorical analysis helps us more fully understand how both, and all, of these explanations hold some truth. As noted more recently by feminist theorist Elizabeth A. Wilson, "the hormone is from the beginning a biologically impure object." Paraphrasing the work of feminist critics of biology such as Anne Fausto-Sterling, Keller, and Donna Haraway, Wilson reminds us that "there are no entities or events . . . that can legitimately lay claim to being biological and not also cultural or economic or psychological or historical."[23] Thus, as Wilson's comment illuminates, when we view, from a topological perspective, the long rhetorical history that connects hysteria to hormones, the debate between Oudshoorn and Roberts becomes, in some ways, irrelevant. Topology helps us to understand that a key function of science is to preserve ideas from the past, to prevent ruptures that would otherwise threaten the existence of the smooth topological space.

The persistence of ancient beliefs about hysteria makes clear that a linear narrative of scientific progress is an incomplete way to tell the story of what has occurred in the change of beliefs between ancient notions of hysteria and those that persisted into the late twentieth century. Changes have occurred in our understanding of these phenomena not only because of new scientific discoveries, but also because of a series of ruptures that have been on the verge of occurring; as one understanding comes to be discordant with prevailing social and scientific norms, there is pressure on science and society to develop a new understanding or conceptual framework. By understanding the job of science as topology, we can see that its task is both to produce the kind of new ideas that create ruptures with old beliefs and to prevent such ruptures from occurring. If those ruptures were to occur, the surface would no longer be topological—namely, it could no longer appear to be smooth and continuous. Thus, the work of science comes to be seen in a new light—that is, even though it seems as though science always looks ahead or always moves forward, we have to acknowledge that another important element of the rhetorical movement that sustains science is devoted to preserving the smooth surface of old ideas, thereby reshaping old concepts and ideas in a way that makes them seem new enough to meet the demands of the present, but without effecting a complete break from the past. These new understandings or frameworks are typically touted as more scientific and trustworthy than their predecessors were, but this does not mean that they escape the confines of the present time and place; rather, if the new ideas seem more logical than the old do, that is at least in part because they correspond more neatly (or relate in a more seamless manner) with the surrounding social and cultural influences of the moment.

It is always assumed in the language that accounts for sex difference that women's knowledge and ways of coming to that knowledge must be different from men's knowledge and ways of coming to that knowledge. Over the several centuries that this book examines, when different groups of experts have vied to offer the most persuasive explanation—"mansplaining" at its finest—each of these explanations has been, in reality, part of a consistent effort to control knowledge. At no time has it been acceptable or desirable for women to have knowledge about their own bodies. Virginity has historically been attested to from the outside, based on someone else's viewing of evidence and not on a woman's own report; hysteria has to be medically explained, or attributed to another factor, such as witchcraft. An important consequence of these assumptions is that, even in the most recent scientific literature, it is often suggested that women need medical or legal experts (who have been, until recently, mostly male) to explain and help them cope with those aspects of their bodies that make them female. Viewed from this perspective, the history of hormones is a history that is permeated by several key ideas that continue to resurface, taking on new shapes and forms to meet the exigencies of each new era. These ideas never really go away, but they continually change as they come to be expressed in the new language of each era, and by the new group of experts that comes to monopolize specific health conditions in different historical periods.

Historians of medicine who have studied this topic tend to echo the narrative of scientific progress that is usually offered in medical texts to characterize the dramatic shifts in belief that we see between ancient times and the present, with different eras espousing different beliefs, and each one more scientifically grounded and, thus, better than the previous belief set. Thus, previous historical analyses offer important insights into both hysteria and hormones, but because they address the two topics separately, none of these analyses has considered the precise series of rhetorical events and strategies through which hormones gradually replaced hysteria as the explanation for female problems. In addition, a common theme in previous historical analyses of hysteria and hormones is that a major shift in thinking occurred in the nineteenth century, when a scientific mindset came to replace a religious or moral understanding of subjects such as sex difference. This tendency to focus on the nineteenth century as an important turning point may also explain why the two topics that are addressed in this book—hysteria and hormones—have been addressed separately but not together. To see hysteria and hormones as continuous with each other requires rejecting long-standing assumptions about the nineteenth century as a time that was defined by major changes in the scientific and public mindset. Instead of working from that assumption, this book's rhetorical analysis looks inside of the nineteenth century, exposing this period as a time of turmoil, conflict, and strife among the numerous epistemological frameworks that competed at the time to explain topics such as hysteria. This move enables us to see hysteria and hormones as fundamentally implicated in each other—as deeply connected and continuous—rather than as distinct explanations that can be differentiated by classifying one concept— hysteria—as mythical and the other—hormones—as scientific.

Multiple Shifts in Perspective

I hope that this new understanding of hysteria, hormones, and the relationship between them stands on its own as an important contribution to the scholarly literature about the history of women's health. But rhetorical analysis, at its best, should accomplish multiple shifts, disruptions, and reorientations. After a good rhetorical analysis, the world should never look the same as it did before, and I would hope that such a changed perspective extends beyond the confines of academic discourse in history, rhetoric, and philosophy of science. As I have suggested in chapter 4, one of the shifts that can occur when we understand time as topological is that discrete points that previously seemed far apart can suddenly become close together. Reflecting this shift in perspective, I have devoted most of this book to collapsing the apparent distance between hysteria and hormones, intentionally bringing these seemingly disparate constructs closer together. In the pages that remain, however, I address other forms of collaps-ing—namely, collapsing between scientific and political female embodiment, between biological and cultural constructs of gender, and between scientific and humanistic ways of knowing—that I hope will facilitate additional forms of shifting, disruption, and reorientation for the readers of this book.

The Science and Politics of Women's Bodies

In a world in which women's rights to control their bodies are increasingly under attack from conservative legislators and politicians who are reviving ancient fears of everything female, questions about the connections between old, obviously misogynistic rhetorical understandings of women's brains and bodies and the newer, ostensibly progressive rhetorical understandings take on increasing signifi-cance. As I have said in chapter 5, when a politician such as Akin, who possesses no substantial knowledge of female matters or female biology, attests publicly that a woman cannot become pregnant if she is "legitimately" raped,[24] it might be easy to dismiss those remarks as coming from a crackpot politician who has no credi-bility as a scientist. And, of course, we might take some solace from the fact that Akin lost his election after those remarks went viral. But now that we are living in a new reality, we cannot afford to feel quite so comfortable about President Don-ald Trump's well-documented history of public misogynistic remarks. It might have seemed pretty bad when, in 2015, the most notable of these remarks was a derogatory statement he made about a Fox News host whom he suggested had been hard on him because she was menstruating,[25] and his statement that if we make abortion illegal in the United States, then women who have abortions will need to be punished.[26] If only that had been the end of Trump's contributions to twenty-first-century rhetorical constructions of the female body.

As I have argued, there is a danger in setting aside outrageous remarks such as these—whether they are made by a crackpot misogynistic lunatic who *loses* an election after making the remarks, or by a crackpot misogynistic lunatic who *wins* an election and becomes president of the United States after making these remarks. Trump especially seems to have the rhetorical effect of being so ridiculous that even those who are most opposed to his political viewpoints can sometimes laugh at his remarks and dismiss them as harmless. But events in the real world, and especially legislative activity at the state level, are another powerful reminder that remarks like these are never just "mere rhetoric," existing on the fringes of society without any ability to harm women. Rather, they indicate deeper misogynistic tendencies that pervade our current daily lives. Most notable among these issues currently is reproductive rights. Although the *Roe v. Wade* Supreme Court decision established in 1973 that women's rights to control their bodies by terminating unwanted pregnancies superseded any imagined right of embryos or fetuses involved in any pregnancy, this decision has since been under attack by right-wing conservatives. The effect of such attacks was relatively weak throughout much of the late twentieth and early twenty-first centuries. Between 2011 and 2013, however, the United States witnessed heightened antiabortion fervor. In fact, during this three-year period, more state-level abortion restrictions were enacted than during the entire decade prior to that.[27]

As I started to draft this chapter in early June 2016, the US Supreme Court was deciding whether the state of Texas should be allowed to uphold a law passed by the state legislature in 2013 that had temporarily closed all but thirteen of the state's abortion clinics. Thankfully, on June 27, 2016, the Court announced its decision that this law could not be upheld because it posed an undue burden that restricted women's access to abortion.[28] If the Supreme Court had upheld this law, all but eight of the state's clinics would have been forced to shut down. Even in the situation as it existed at that time, women in some areas of the state had to travel more than 200 miles to the nearest abortion clinic—a situation that has proven difficult to change even after the Supreme Court's decision.[29] And once a woman arrives at a clinic in Texas to seek an abortion, she must still be subjected to a condescending set of laws that require her to hear state-mandated counseling information and then undergo an ultrasound in which she is forced to listen to the fetal heartbeat, watch the ultrasound images on the screen, and listen to the physician's explanation of the images—all before waiting twenty-four hours to return to the clinic to have the procedure.

All of these rhetorical moves are supposedly intended to protect the woman's health and safety.

Meanwhile, in another rhetorical situation in Texas that might seem entirely separate, adults and children are suffering real health problems because of state legislators who refuse to interfere with the oil and gas industries' rights to pursue profit at all costs. When it comes time to consider implementing regulations that would ensure citizens' safety from environmental and physical risks posed by the practices of the oil and gas industry, where are these state legislators who are so concerned about women's health and safety, about preserving "life" at all costs? And, if we are going to focus on fetuses, who is looking out for the women who suffer miscarriages of wanted and intended pregnancies because of chemicals that we allow in our environment in the name of industry and free enterprise?[30] The salient point of contrast between these two different rhetorical situations is that the people (many of whom are men) who run these oil and gas corporations are assumed from the outset to be rational beings with strong minds, capable of running their companies in a manner that will protect the safety and health of Texas citizens. In fact, self-reporting has been the usual method of ensuring industry compliance with guidelines established by the Environmental Protection Agency and other regulating bodies—even when people living in affected areas suffer devastating health effects as a direct result of industry practices.[31] By contrast, the women who might seek abortions in Texas are not seen as having that same capacity to self-regulate, reflect, and make their own rational decisions; rather, they are rhetorically positioned as less-than-competent individuals who need the government—through the mouthpiece of physicians—to provide them with information and guidance to ensure that they will make the right decision.

These may seem like disparate issues, but they are not. If a rhetorical analysis is to be truly transdisciplinary, then scholars of science, medicine, and rhetoric must reach out and identify connections among different domains that may not appear to be related. As I suggest throughout this book, one of the ways in which rhetorical scholars might respond to the reimagining of time and space that emerges through Serres's topology is by dramatically altering how we understand the scope and scale of our work as rhetorical critics. Through such an alteration, objects or events that seem far apart can suddenly seem close together in a way that is similar to what happens to specific locations on the surface of bread dough when it is kneaded, or in the folding and unfolding of a handkerchief, as Serres observes. The act of placing close together the seemingly disparate examples and rhetoric that I have juxtaposed above—that is, placing rhetorics that govern the regulation of the female body alongside rhetorics that govern the regulation of powerful industries such as those of US oil and gas production—reveals logical inconsistencies that are quite striking and would not otherwise be apparent. In particular, this juxtaposition of two points that would otherwise seem to exist far apart from each other reveals a fundamental difference between our society's understandings of female rationality and male rationality. The only possible way to explain this difference is by acknowledging the phenomenon that manifests itself in a virtually endless number of rhetorically produced scientific configurations throughout the eras, as examined in this book, in which the female mind and body emerge as foreign, mysterious, or defective versions of the male mind and body.

As I have argued elsewhere, these configurations of the female mind and body in contrast to those of males are even embedded in the language of the 1973 *Roe v. Wade* Supreme Court decision. The justices who crafted that decision did not know what to do with the pregnant body, so they conceptualized it in a way that allowed for an ambiguity and uncertainty that corresponds with the rhetorical configurations of femaleness[32] that have been foregrounded in this book. Even in that text, which was a landmark for twentieth-century women's rights, women's bodies are construed as dark, unknown territory. In the case of *Roe v. Wade*, that situation happened to work in favor of abortion rights because the justices grounded their interpretation in favor of women's reproductive rights on an acknowledgment that we cannot know exactly what is happening inside a woman's body when she is pregnant. But that same ambiguity can be easily turned around and used against women, which is what we are seeing now with the resurgence of so many antiabortion regulations. In fact, in the arguments of today's conservative politicians, this same ambiguity about what goes on inside a woman's body is used to achieve different rhetorical ends than what we have seen in the arguments of politicians in the past. In particular, we can see this ambiguity being employed in two different ways in today's rhetorics: politicians claim, in some cases, that they know what happens inside the pregnant body (i.e., that they are certain that the fetus counts as a person with the same rights as the adult woman who carries it) or, in other cases, that even if we cannot know this for certain—as in the case of fertilized but unimplanted eggs—to give the benefit of the doubt to the state and its interests rather than to the woman herself. In neither of these rhetorical configurations is the woman construed as an individual, whole human being with fundamental rights to interpret with her own mind what occurs inside her body. It is assumed in these different rhetorical configurations that women's knowledge and ways of coming to that knowledge must be fundamentally different from men's knowledge and ways of coming to that knowledge, and it is not a coincidence that women's knowledge is mistrusted and feared, whereas men's knowledge is treated as authoritative, rational, and necessary.

Recall that when Eve bit the apple that the book of Genesis describes as "desirable for gaining wisdom," the wisdom that she acquired did not make her smarter than Adam. The knowledge of that very first woman could not be trusted on its own terms as valuable or accurate; instead, it was said to lead to the irreversible downfall of the human race. And in today's scientific rhetoric, we see some of these same ideas resurface when scientists scrutinize the loss of mental capacities that occur when female rats or humans are pregnant, or when they study how women's stress responses differ across the phases of the menstrual cycle. The implication of such research is that women, even if they are normal and healthy, need to consult the experts to help them manage their complicated bodies as they proceed throughout the various phases of their lives.

Biological and Cultural Constructs of Gender

When we shift our focus from looking at the scientific rhetorics that explain how hormones function in the female body to looking at actual lived experiences of women in today's world, we can collapse some of the distinctions that we usually assume to exist between the biological and social aspects of gender. On one hand, we see the recurring, anecdotal reports of pregnancy or mommy brain, as discussed in chapter 7, in which women have declared their incompetence to handle professional and intellectual work after having babies. But on the other hand, some current workplace research reports the experiences of women who take extreme pride in their ability to handle both roles. For example, according to a recent interview study by Patrice M. Buzzanell, many participants seem to have internalized the ways of thinking about their own minds and bodies that are embedded in scientific studies of female hormones,[33] which I analyze in this book. Many of these women intentionally strive to adhere to the impossible ideals that are created for women in contemporary scientific rhetorics, which I have examined. These rhetorics depict women—their bodies and brains, and the relationship between their bodies and brains—as problematic or as having distorted versions of men's bodies and brains. However, in the most recent scientific rhetorics, as I have argued, the experts seem to allow for the possibility that women can take extreme measures to overcome their inherent deficiencies by working very hard to have the appropriate response to the adversities that they encounter in the home or workplace or, in some cases, by taking the appropriate medication to restore chemical or hormonal balance.[34] If we conceptualize contemporary rhetorics of the female body as vast topological spaces, it makes sense that these biological and social aspects could collapse and suddenly appear to be much closer to each other than they initially seemed. As I have suggested in chapter 7, the many enthymematic arguments, both scientific and public, that are enabled by our collective faith in hormones as the cause of all of women's problems is one important factor that allows this collapsing to occur.

Buzzanell's study in particular includes several examples to illustrate how women internalize the notion that they are expected to adhere to some unstated yet very real notion of what it means to overcome the inherent difficulties of being a woman, whether those difficulties are defined as some distinctive feature of the female brain and how it has to adapt to hormones, or whether those difficulties are defined as existing in the real world alongside what women have to do to negotiate the demands of combining childcare with professional success. One powerful example in Buzzanell's study is "Julie," who, "while proud of her role as arranger, does reflect on the amount of thinking that she did over the years about managing children and career." Julie realizes that "this thought and planning is part of doing gender."[35] Like Julie, many mothers interviewed in Buzzanell's study reported that they have had to engage in this kind of "sensemaking" in order to integrate their successes as mothers and working professionals. The "sensemaking" that Buzzanell and her coauthors describe can be seen as a further example of women trying to shape themselves—that is, to gain control over their emotions, rearrange their family lives and childcare arrangements, and so on—to fit the needs of society, rather than expecting society—or their family, coworkers, or employers—to change to accommodate their needs. Another important theme in Buzzanell's study is that when women described the reasons why they work, they almost always mentioned someone else. It is for the economic wellbeing of their family, and it fits "consumerist patterns," as Buzzanell says.[36]

Similar themes appear in the most recent scientific literature that proclaims to offer a more positive interpretation of the cognitive effects of pregnancy hormones. As I have noted, these cognitive benefits are often portrayed as something that mothers need to pursue actively; this contrasts with "pregnancy brain," which is depicted as a negative effect that is a biological inevitability. Along these lines, Thornton examines how neuroscientific discourses are reflected in popular parenting advice books and, specifically, how mothers are urged to engage in activities that maximize their children's brain capacities. These new advice manuals, according to Thornton, emphasize the freedom that comes with good mothering—the freedom to relax and enjoy the baby rather than doing everything in one's power to maximize the baby's neural capacity—and the fact that good mothering is supposed to be enjoyable. These "back-to-basics discourses," according to Thornton, "reinvigorate and adapt these themes in ways that articulate with late capitalism's demands for flexible, entrepreneurial subjects." According to Thornton, these discourses are typical of neoliberalism, as described by Foucault, because they emphasize freedom and individual choice while, at the same time, subjecting mothers to increased scrutiny through "ever-more-dense networks of authority, expertise, and government." Furthermore, maternal emotion becomes "a calculable space for surveillance and intervention." Thornton mentions such examples as the "first three years" movement in the 1990s, which is exemplified by the *Baby Einstein* series of products. She notes that the recent back-to-basics rhetoric is grounded in the same neuroscientific ways of thinking, but she notes a backlash against the corporate involvement of that earlier movement. These "new" discourses revive old ideas, and a topological analysis helps illuminate the fact that "the back-to-basics discourses rearticulate and reenergize the historically persistent tie between women and emotion in a distinctly neoliberal affective context circumscribed by notions of choice, identity, and freedom."[37]

As Thornton says, neuroscience is a popular explanation right now because of its great "translatability." Through powerful persuasive devices such as digital images of brain activity, popular neuroscience convinces us that we must continually monitor our emotions, and that negative ideas or thoughts can be eliminated through the proper behaviors. Mothers, in this way of thinking, are responsible for attaining their own freedom through the enjoyment of the motherhood experience because "attachment is a project of authenticity that requires women shape themselves into mothers who *genuinely* enjoy the early experiences with their infants."[38] This neuroscientific turn might be seen as the latest phase in a long tradition of rhetorics that make women responsible for their own as well as society's problems.[39]

Some of Thornton's observations resonate with the findings of Buzzanell's research, which showed that mothers in the workplace often internalize these conflicting rhetorics and that they do so in ways that are not always empower-ing. Women in these situations take it on themselves to produce a positive response to the most negative external conditions. What if we flipped this situ-ation around and placed our emphasis on fixing such heinous external condi-tions, rather than on fixing women by correcting their hormones, emotions, or behaviors? An important theme that emerges in both Thornton's and Buz-zanell's studies is the many ways in which mothers internalize the same rhetori-cal configurations that are articulated in scientific studies—namely, that there are positive and negative ways to respond to situations of adversity, regardless of whether those situations are created by a difficult workplace environment or by hormones inside women's own bodies, and that individual women are respon-sible for cultivating positive responses.

Previous feminist scholars have already made moves in this direction. For example, suggestions for alternative explanations of the twenty-first-century hormonal woman can be found in the burgeoning field of feminist psychology. These approaches are not directly opposed to the biological explanations of hormones' effects on the female brain that have persisted into the twenty-first century, and they do not completely ignore the existence of biology-based sex difference. For instance, Ussher advocates for finding a balance between social and biological explanations of gender, noting the deficiencies in each of these when they are used as stand-alone approaches. In considering these feminist-psychology approaches, however, it is important to note how these feminist scholars discuss gender difference, in contrast to the strictly biological approach that defined women's behaviors and mental capacities only in terms of their hor-mones. Ussher, for instance, cites a study that suggests a more nuanced under-standing of the relationship between biological and social factors in women's responses to stress. This study suggests that "in the face of stress, women are more likely to use coping strategies that involve verbal expression to others or the self—seeking emotional support, rumination, and positive self-talk—whereas, in contrast, men are said to engage in avoidance in the face of stressors that involve relationships or other people." Ussher cites additional psychology research that suggests that "women are deemed to self-silence because they believe that they are not loved for who they are, but for how well they meet the needs of others, with the result silencing of needs and anger, and the use of external stan-dards to judge the self, leading to feelings of worthlessness and hopelessness."[40]

Ussher contends that even with all the variation in these alternative theories from psychology, problems persist in understanding this situation from a feminist perspective. She contends that these explanations "still position depression within the woman," and in so doing, they obscure the gendered divisions of domestic labor, disparities in pay, prevalence of sexual abuse, and lack of reproductive rights for women that still exist in our society.[41] Ussher ultimately calls for a more complex, multidimensional model: "Examining the construction and treatment of depression also provides insights into the cultural construction of what it means to be 'woman,' where diagnosis with pathology is an ever present spectre, whether we accept or reject archetypal feminine roles."[42]

But an important part of Ussher's message is also that women, individually and collectively, have the capacity to resist these deeply embedded cultural constructs—even those constructs that operate as widely accepted scientific truths. She offers several suggestions for rethinking these long-standing constructs about "raging hormones" and the like. For instance, she observes, one strategy that individual women can use is to "reject idealized fantasies of femininity which tell us that we must be calm, controlled and able to cope at all times," and at the same time, to resist such tendencies as overcommitment and self-sacrifice that can intensify the anger and distress that are characterized as the "monster within" women.[43]

Along similar lines, Roberts argues for a "refigured view of hormones as messengers of sex," and she suggests that "hormones do not message an inherent or preexisting sex within bodies, but rather are active agents in bio-social systems that constitute material-semiotic entities known as sex."[44] Roberts concludes by arguing for a less deterministic way to think about the manner in which hormones operate in the body. She cites the etymology of the word *hormone*, which means "to excite" or "to provoke," and she suggests that this meaning might open the way for a more productive, interactive way to think about the kind of communication inside the body in which hormones are involved. This communication can be conceived of less like a telegram and more like the multiple forms of communication that are available in the twenty-first century. Along similar lines, feminist theorist Sari Irni argues for a "posthumanist performative approach to sex hormones."[45] Irni extends the earlier work of Roberts and others by arguing that "a feminist approach to sex hormones needs to account for the context-specific power relations within which many of the existing 'hormonal changes' come to materialize."[46] She also argues that we must go beyond just technoscience and medicine to understand contextual apparatuses such as the welfare state and how such an apparatus can contribute to the context in which hormones are regulated and meanings of hormones are produced. These efforts to develop a feminist study of hormones might stand as a powerful contribution to what Elizabeth A. Wilson calls "'gut feminism'—a feminist theory that is able to think innovatively and organically at the same time." In referring to the gut, Wilson engages in just the kind of collapsing between biological and cultural that would be necessary to achieve a feminist understanding of hormones. As Wilson reminds us, "the gut is an organ of mind: it ruminates, deliberates, comprehends."[47] Intentionally reversing the feminist tendency in recent decades to eschew the biological elements of gender, Wilson then encourages feminist scholars to develop a "conceptual toolkit for reading biology."[48]

It seems possible that popular beliefs might possibly evolve in directions that align with some of these new rhetorical reconfigurations of hormones that are suggested in recent feminist scholarship. One example is a popular book that was published in early 2016. Written by physician Judy Holland, the book *Moody Bitches* encourages women to reclaim the hormonal flux that the Western world has taught them for so many centuries to deny and suppress. In Holland's words, "our hormonal variations allow us to be empathic and intuitive—to our environment, to our children's needs, and to our partners' intentions. Women's emotionality is normal. It is a sign of health, not disease, and it is our single biggest asset. Yet one in four American women are choosing to medicate away their emotionality with psychiatric medications, and the effects are more far-reaching than most women realize."[49] Although some of Holland's advice runs the risk of placing increased blame and responsibility on individual women rather than addressing the societal structures that continue to perpetuate gender inequities, her book is valuable in that it could potentially reach wide audiences with its message that women are not fundamentally pathological—that we are not imperfect, flawed, or disordered versions of men, either in our physical or mental selves.

Sciences and Humanities: From Explaining to Exploring

"It was on a recent trip to Indonesia," writes *New York Times* reporter Julia Baird, "that, as a male bureaucrat sounded forth on a vast span of subjects without being asked to do so, I realized that the English language was in need of a new addition: the manologue." Baird defines this new term as having "many forms," but being fundamentally "characterized by the proffering of words not asked for, of views not solicited and of arguments unsought."[50] In many ways, Baird's new term could be said to characterize the long history that has been the subject of this book; the centuries-long history that connects hysteria to hormones might be understood as an endless "manologue" in which men have "sounded forth on a vast span of subjects" in an ongoing effort to explain women to themselves. For a full understanding of this "manologue," my analysis had to reach back to the most ancient texts about hysteria, although the book's main focus is on the rhetorical and scientific events that came immediately before and after the 1905 emergence of the term *hormone* in a lecture by British physician Ernest Henry Starling.

What is the problem with this manologue? Should it be rejected entirely? Should we start from scratch? Would that even be possible? Instead of answering these questions, I would like to close with a call for fundamental changes to the rhetorical situation that surrounds the production of scientific knowledge about the biology of the sexed body. In recent decades, women have had a greater presence in STEM fields. But we are a long way from achieving true equality or anything close to gender equity in scientific knowledge production, especially with regard to the biology of the sexed body.

I believe that recent research in the new field of social physics has potential to reorient this as a situation in which our whole world can benefit from achieving a knowledge-producing enterprise that is more inclusive—that is, from achieving an apparatus of scientific knowledge production that incorporates contributions from a wider, more diverse group of knowledge producers. Alex Pentland's book *Social Physics: How Social Networks Can Make Us Smarter* uses an approach that Pentland calls social physics to provide quantitative evidence that organizations function more effectively—and are more productive and successful—when mechanisms are in place to ensure that ideas are, as Pentland says, harvested from everyone in the organization. Pentland defines social physics as "a quantitative social science that describes reliable, mathematical connections between information and idea flow on the one hand and people's behavior on the other." He and his team use this new science to provide empirical evidence that demonstrates the value of achieving broad input in decision-making from all sectors of an organization, rather than limiting decisions to a few individuals who are located at the top of an institutional hierarchy. In Pentland's words, social physics "enables us to predict the productivity of small groups, of departments within companies, and even of entire cities. It also helps us tune communication networks so that we can reliably make better decisions and become more productive." As Pentland then says, "when decision making falls to those best situated to make the decision rather than those with the highest rank, the resulting organization is far more robust and resistant to disruption."[51] Working with his team of graduate students and colleagues in the MIT lab that he directs, Pentland designed a method of collecting data on various human interactions within specific organizations, including electronic communication such as email and precise counts of the quantity and nature of face-to-face interactions and phone calls. He claims that this groundbreaking method of data collection provides quantifiable evidence of the monetary value that an organization can accrue by achieving a more diverse workforce and making sure that everyone in this workforce is able to contribute good ideas.

Although Pentland's social physics approach is implemented in the context of specific organizations and is thus geared toward business professionals, I believe that the ideas established in his team's study can potentially revolutionize the arguments that we make in favor of diversity in STEM fields and in the academy more broadly. Rather than continuing to depict this as a problem that individuals face and that institutions need to solve to benefit these individuals, we can understand diversity as a goal that will enable academic institutions— and the scientific enterprise at large—to produce more and better knowledge.

In the *New York Times* article in which Baird coins the term *manologue*, Baird discusses alternative approaches that usefully illuminate the large-scale change to the scientific knowledge enterprise that I am calling for: "Women use words to explore, men to explain."[52] This distinction between explaining and exploring is a fascinating one and could be used to move forward and achieve a better, more inclusive understanding of many of the concepts and issues that this book addresses. In many ways, my book could have been titled "Centuries of Men Explaining Women to Themselves." Explaining has a power-over dynamic that is not present in exploring, which implies a much more open-ended outcome. In *Conversations with Latour*, Serres speculates about a new way to conceive of the light that is shed through academic inquiry, using language that resonates with what I suggest: "No, here we're not talking about the light of a platonist sun, nor that of the Aufklärung—so purely physical that it blinds us toward the social sciences—nor is it a question of distinguishing, since we are trying to understand the famous link. Rather, we are talking about a fairly soft and filtered light that allows us better to see things in relief, through the effects of contrast produced by rays and shadows that melt together, that are mixed, nuanced."[53] The present volume, *From Hysteria to Hormones*, could itself be construed as an attempt to find an alternative mode of knowledge production. Because this is a critical exploration, my goal has been not to provide definitive answers—to respond to "mansplaining" with my own version of "womansplaining"—but rather to point to possible new directions.

Despite the significant changes in understandings of women's health and physiology that occurred between ancient Greece and the present, the history that I have examined in this book is permeated with long-held Western beliefs that posit a divide between the mind and the body and that assign gendered values to this dichotomy. This set of beliefs values reason (affiliated with the mind) over emotion (affiliated with bodily urges and impulses); it affiliates the former with masculinity, and the latter with femininity, and it has been reshaped throughout the centuries in response to different social and political beliefs and historical events that predominate at different points in time. Another important theme is that each new set of experts who have come along to offer new explanations for this phenomenon has done so with the suggestion that they are offering better evidence, more empirically grounded, more scientific; no one among those whose ideas I have analyzed claimed to be going backwards to a more religious, less scientific explanation. And it is always women's bodies, women's symptoms—not men's bodies or men's symptoms—that are categorized as problematic or as in need of diagnosis or explanation. Another consistent theme is that of control. Although the causes to which a lack of control are attributed have changed dramatically over time, the desire to control women's bodies—to perceive women's bodies as less controllable than men's bodies, and to express a desire to reign in their inherently out-of-control state—has remained constant. Also important in this persistent mindset is the notion that women experience a divide between their rational selves and their nonrational selves that is fundamentally different from what men experience; the suggestion is that women are driven by something invisible inside them that makes them act and think in certain ways that are beyond their own rational control, and that the relationship between the self and the body is, for women, much more complicated than it is for men. One final important theme that becomes especially apparent in current discourses on women's reproductive rights is the implication that by solving women's health problems, we can solve society's problems as well; this theme is sometimes reversed, depending on the historical era; the logic in this case is that solving society's problems will also solve women's health problems.

If we adopt a social-physics mindset, the problem with this long history is not just in the content of the ideas that it has perpetuated, but also in the fact that the knowledge production has been almost entirely a one-way process, with men producing knowledge about women. That one-way process of knowledge production, accumulated over so many centuries, is why even in today's scientific discourse we see so many deeply embedded judgements that are made by men about women. If it is true we live in a world in which there has been a centuries-long tradition that dictates that one group will be the knowledge-producers and another group will be the object of the knowledge that is produced, then it is not surprising that the latter group has and will continue to be perpetually depicted as mysterious, pathological, uncontrollable, and in need of further explanation.

Insofar as this book is a transdisciplinary analysis of rhetoric, I hope that these suggestions about reorienting our entire approach to the scientific knowledge-producing enterprise have value not only for humanities scholars, but also for scientists. For the most pragmatic reasons, in a world in which competition for grant funding and other resources that support the scientific enterprise are increasingly intense, paying attention to the series of rhetorical events that enable a new concept to ignite the scientific community—such as that which occurred when Starling first used the word *hormone*—has a value that is hard to quantify but should be acknowledged.

Rhetoric, in all its worldliness, its negative connotations, its imperfections and unpredictability, is as important today as it was in Aristotle's world. As the number of channels for producing and communicating new scientific knowledge continue to multiply, and as the audiences for such knowledge become more complex, learning what it takes for a new concept or discovery to be known as the breakthrough that forever changed scientific understandings of an aspect of the natural world or the human body makes sense. This ability to use language and multiple media forms to communicate will increasingly be what distinguishes the excellent STEM professional from the one who just gets by, and there is no better way to gain this skill set than by studying rhetoric. If we can train a wider, more diverse population than ever before to participate in this knowledge-producing endeavor, all of us will ultimately have a better world to live in.

Although we have made some progress in ensuring that this broad set of communication skills is available to a wider, more diverse group of stakeholders, we only need look at recent issues of the *Chronicle of Higher Education* or other similar publications to learn the latest depressing statistics that tell the story of the academy's failure to achieve racial and gender equity at the highest ranks. Women and men graduate from PhD programs in relatively equal numbers now, but as we proceed up the ranks in the university, the numbers of women dwindle, drastically and sadly.[54] Even when such articles are written by authors with the most progressive mindset, these problems of diversity, or lack thereof, are usually depicted as problems that are confronted by individuals, and even when institutional changes are suggested, the anticipated outcomes are usually presented as outcomes that will benefit individuals—helping women and minorities achieve greater career success, helping them break the glass ceiling, and so on. The approach that I am calling for would revolutionize how we frame these arguments about diversity. The approach I am calling for would require that our starting place in any higher education diversity initiative must be to agree that it is better for the institution and the world if every academic unit that produces knowledge in every university were truly reflective of diverse perspectives. This would not entail checking off boxes on an Equal Employment Opportunity form; it would require actively seeking to include individuals who reflect different perspectives about the world, individuals who are biologically, intellectually, geographically, socially, and spiritually diverse. It would involve going out of our way to ensure that these individuals have every opportunity to advance to leadership positions and be represented at every rank and location in the university. And it would demand doing all of this not only because it looks good, not only because it is fair or beneficial to these individuals, but because it leads to the production of knowledge that is of better quality by every possible standard.

The story told in this book is a story about things that change and things that do not change, ever. In many respects, the story that I have told could be seen as depressing or sad; it is a story that does not necessarily reinforce our faith in the ability of scientific progress to free us from the oppressive beliefs of the past. But I have tried to close with some more hopeful suggestions about how we might reconfigure not only ourselves but the entire knowledge-producing apparatus that is academic science and medicine in the twenty-first century. In the end, we will be better off if we can achieve a world in which knowledge production is not a one-way street but rather a smoothly functioning freeway system that is designed to make sure that all people can get to their intended locations as independently, efficiently, and safely as possible. Let's replace the "manologue" with a conversation that is more democratic, more inclusive, more transdisciplinary, and that will ultimately lead to better scientific knowledge and a better world.

Notes

Preface

1. Gen. 3:6.
2. Demand, *Birth*, 32.
3. Ibid., 36.
4. Ng, "Hysteria," 288.
5. Tasca et al., "Women."
6. Starling, "Croonian."
7. J. Wilson, "Charles-Edouard Brown-Séquard."
8. See, for example, Oudshoorn, *Beyond*; Martin, *Woman*; and Roberts, *Messengers*.
9. Serres and Latour, *Conversations*.
10. American Psychiatric Association, *Diagnostic, Fourth Edition*.
11. DeBondt et al., "Regional."
12. Kinsley and Lambert, "Reproduction-Induced."
13. Albert, Pruessner, and Newhouse, "Estradiol."
14. WebMD LLC, "Escape."
15. InfoSpace Holdings LLC, "10 Things."

Chapter 1

1. "Topology."
2. Serres and Latour, *Conversations*, 58–59.
3. Demand, *Birth*.
4. Oudshoorn, *Beyond*, 150.
5. Martin, *Woman*, 41.
6. Roberts, *Messengers*, 21.
7. Jensen, "Improving," 331.
8. Ibid., 332.
9. Pettit, "Becoming," 1052–54.
10. Ibid., 1075.
11. Roberts, *Messengers*, 46.
12. Jensen, "Improving."
13. Hausman, "Ovaries."
14. Segal, *Health*, 22.
15. Ibid., 36.
16. Keränen, *Scientific*.
17. Emmons, *Black*.
18. Teston, "Grounded."
19. Derkatch, *Bounding*.
20. Seigel, *Rhetoric*.

21. Brown, "Michel Serres," 9.
22. Serres, *History*, quoted in Brown, "Michel Serres," 9.
23. Rickert, *Ambient*; Eskin, "Hippocrates."
24. Ballif, "Writing," 246.
25. Harris, *Rhetoric*.
26. Campbell, "Anxiety"; Campbell, "Scientific."
27. Condit, *Meanings*.
28. Serres and Latour, *Conversations*, 148.
29. Condit, "Mind," 3–4.
30. Foucault, *Birth*.
31. Segal, *Health*, 26.
32. Happe, *Material*.
33. Veith, *Hysteria*; Demand, *Birth*.
34. Thompson, *Wandering*; Maines, *Technology*.

Chapter 2

1. Feudtner, *Bittersweet*.
2. Segal, *Health*.
3. Kotchen, "Historical."
4. American Psychiatric Association, *Diagnostic, Fourth Edition*.
5. Veith, *Hysteria*.
6. Slater, "Diagnosis," 1395.
7. Latour, "Enlightenment," 83.
8. Ibid., 88.
9. Serres and Latour, *Conversations*, 145.
10. Miller, "Aristotelian," 130–31.
11. Serres and Latour, *Conversations*, 148.
12. Fissell, "Hairy."
13. Bullough, "Early."
14. Fissell, "Hairy," 43.
15. Bullough, "Early," 236–37.
16. Hollick, *Diseases*.
17. Hollick, *Origin*.
18. Ng, "Hysteria," 288; Demand, *Birth*, 32; Thompson, *Wandering*; Veith, *Hysteria*.
19. Thompson, *Wandering*, 31–33; Veith, *Hysteria*, 2–6.
20. Ng, "Hysteria," 291.
21. Ibid.
22. Ibid.
23. Ibid.
24. Ibid., 292–93.
25. Demand, *Birth*, 55.
26. Merskey and Merskey, "Hysteria."
27. Ibid., 401.
28. Ibid.
29. Ibid., 402.
30. Boss, "Seventeenth-Century," 222.
31. Ibid.
32. Tasca et al., "Women."

33. Boss, "Seventeenth-Century"; Merskey and Merskey, "Hysteria"; Veith, *Hysteria*; Tasca et al., "Women."

34. Boss, "Seventeenth-Century," 224.

35. Ibid., 221.

36. Ibid., 226.

37. Ng, "Hysteria," 290.

38. Boss, "Seventeenth-Century," 225–26; see also Zimmer, *Soul*.

39. Thompson, *Wandering*, 135.

40. Veith, *Hysteria*, 6–7.

41. Tasca et al., "Women"; Boss, "Seventeenth-Century," 224–25; Merskey and Merskey, "Hysteria," 402; Veith, *Hysteria*.

42. Boss, "Seventeenth-Century," 222.

43. Shields, "Functionalism," 745.

44. *Aristotle's Masterpiece*, 74.

45. Ibid., 75.

46. Ibid., 94–95.

47. Hollick, *Diseases*, 194–216.

48. Ibid., 195.

49. Hollick, *Origin*, 609.

50. Ibid., 611.

51. *Aristotle's Masterpiece*, 78.

52. Ibid., 74.

53. Hollick, *Origin*, 613.

54. Ibid., 640.

55. Ibid., 641.

56. Ibid., 660–66.

57. Ibid., 679.

58. *Aristotle's Masterpiece*, 36.

59. Ibid., 37.

60. Ibid., 97.

61. Hollick, *Diseases*, 205.

62. Ibid.

63. Ibid., 209.

64. Ibid., 211–12.

65. Ibid., 199.

66. Ibid., 199–200.

67. Ibid., 201.

68. Serres, *Rome*, 82.

69. Connor, "Michel Serres's Milieux."

70. Campbell, "Anxiety"; Campbell, "Scientific."

71. Miller, *Kairos*.

72. Condit, *Meanings*; Condit et al., "Recipes"; Happe, *Material*.

73. Ibid.

74. Emmons, *Black*; Scott, "Public."

Chapter 3

1. Teston and Graham, "*Stasis*."

2. Graham and Herndl, "Talking," 163.

3. Micale, "On the 'Disappearance,'" 497.

4. Prelli, *Rhetoric*, 60.

5. Ibid.

6. Scull, *Hysteria*, 105–7.

7. Gross, Harmon, and Reidy, *Communicating*.

8. Ibid., 30.

9. Ibid., 117.

10. Ibid., 118.

11. Ibid., 119–21.

12. Zimmer, *Soul*; Scull, *Hysteria*, 106.

13. Scull, *Hysteria*, 104–30.

14. Gross, Harmon, and Reidy, *Communicating*, 21.

15. Prelli, *Rhetoric*, 146.

16. Ibid.

17. Ibid., 151.

18. Ibid., 152–53.

19. Veith, *Hysteria*, 156; Scull, *Hysteria*, 25–26.

20. Veith, *Hysteria*, 120–24.

21. MacDonald, *Witchcraft*, xlvii.

22. Boss, "Seventeenth-Century," 225; Veith, *Hysteria*, 128.

23. Veith, *Hysteria*, 132–33. Historians of medicine tend to agree that "animal spirits" in this context resembled the central nervous system more than any other phenomena that we acknowledge in medicine today (Boss, "Seventeenth-Century," 230).

24. Boss, "Seventeenth-Century," 230–31.

25. Veith, *Hysteria*, 136.

26. Ibid., 145.

27. Ibid., 136.

28. Ibid., 175.

29. Ibid., 183.

30. Faber, "Jean-Martin," 284.

31. Ibid., 281.

32. Ibid.

33. Charcot, *Lectures*, 272.

34. Gross, Harmon, and Reidy, *Communicating*, 119–21.

35. Charcot, *Lectures*, 272.

36. Scull, *Hysteria*, 108.

37. Prelli, *Rhetoric*, 148.

38. Scull, *Hysteria*, 107–8.

39. Faber, "Jean-Martin," 277.

40. Ibid.

41. Scull, *Hysteria*, 123; see also Didi-Huberman, *Invention*.

42. Faber, "Jean-Martin," 286.

43. Charcot, *Lectures*, 263.

44. Faber, "Jean-Martin," 279.

45. Ibid., 280.

46. Charcot, *Lectures*, 263.

47. Ibid., 268.

48. Ibid., 268–69.

49. Ibid., 270–71.

50. Ibid., 276–77.
51. Johnson, *Essay*.
52. Faber, "Jean-Martin," 280.
53. Prelli, *Rhetoric*, 149.
54. Schiller, *Möbius*, 28.
55. Faber, "Jean-Martin," 281.
56. Ng, "Hysteria," 290.
57. Faber, "Jean-Martin," 281.
58. Ibid., 283.
59. Ibid., 278–79.
60. Ibid., 276.
61. Charcot, *Lectures*, 300–2.
62. Ibid., 302.
63. Ibid., 306.
64. Bonduelle and Gelfand, "Hysteria."
65. Charcot, *Lectures*, 305.
66. Prelli, *Rhetoric*, 60.
67. Micale, "On the 'Disappearance,'" 499.
68. Ibid., 499–500.
69. Scull, *Hysteria*, 131–33.
70. Micale, "On the 'Disappearance,'" 519.
71. Ibid., 501–2.
72. Ibid., 521.
73. Faber, "Jean-Martin," 288.
74. Scull, *Hysteria*, 109.
75. Bogousslavsky, "Hysteria."
76. Harris, *Rhetoric*, 13.
77. Serres and Latour, *Conversations*, 81.

Chapter 4

1. Serres and Latour, *Conversations*, 60.
2. Jensen, "Improving," 331; Pettit, "Becoming."
3. Lawrence, "Controversial."
4. J. Wilson, "Charles-Edouard," 1403.
5. Ibid., 1405.
6. Lawrence, "Controversial," 1.
7. J. Wilson, "Charles-Edouard," 1405.
8. Lawrence, "Controversial," 2.
9. Henderson, "Ernest," 5.
10. J. Wilson, "Charles-Edouard," 1405.
11. Ibid., 1405.
12. Brown-Séquard, "Effects," 105.
13. Ibid.
14. Ibid., 105–6.
15. Later researchers discovered that the only way testosterone could be extracted in a usable form was from urine, not blood.
16. J. Wilson, "Charles-Edouard," 1406.

17. Lawrence, "Controversial," 3.

18. "Animal Extracts," 1279.

19. Ibid.

20. Ibid.

21. Ibid.

22. Starling, "Croonian," 3.

23. Ibid., 6.

24. Henderson, "Ernest," 9. For an alternative origin story of the term *hormone*, see Bayliss, *Principles*, 712. Bayliss reports that William Bate Hardy, a biologist and food scientist, was the first to suggest the word *hormone* to Starling.

25. Starling, "Croonian," 4.

26. Schiebinger, *Nature's*.

27. Starling, "Croonian," 14.

28. Henderson, "Ernest," 7.

29. Pavlov, "Nobel."

30. Ibid.

31. Ibid.

32. Henderson, "Ernest," 9.

33. Ibid.

34. Jensen, "Improving," 335.

35. Bayliss, *Principles*, 712.

36. Starling, "Croonian," 27.

37. Ibid., 28.

38. Ibid.

39. Ibid., 29–33.

40. Ibid., 35.

41. Vesalius, *On the Workings*.

42. Harvey, *On the Motion*.

43. Gross, Harmon, and Reidy, *Communicating*, 120–21.

44. Ibid., 140.

45. Ibid., 24.

46. Ibid., 142.

47. J. Wilson, "Charles-Edouard," 1407.

48. Ibid.

49. Jensen, "Improving," 332–33.

50. Lawrence, "Controversial," 1.

51. J. Wilson, "Charles-Edouard," 1403.

52. Ibid.

53. Ibid., 1408.

54. Stewart, "Growing," 17.

55. Ibid., 20.

56. Ibid., 23–24.

57. Ibid., 18.

58. Ibid.

59. Gross, Harmon, and Reidy, *Communicating*, 160.

60. Sajous, "Endocrinology," 629.

61. Ibid., 643.

62. Harris, *Rhetoric*.

63. Ceccarelli, *On the Frontier*, 43.

64. Rowntree, "Evaluation," 182.
65. Zimmer, *Soul*, 267.
66. Jensen, "Improving," 331.
67. Starling, "Croonian," 26.

Chapter 5

1. Boylan, "Galenic," 85; see also Fissell, "Hairy," 47.
2. Boylan, "Galenic," 88.
3. Huntley and McClain, "Legitimately."
4. Alter, "Todd Akin."
5. Kliff, "Rep. Todd Akin."
6. Ibid.
7. Serres and Latour, *Conversations*, 48.
8. Stormer, *Sign*, 50.
9. Ibid., 51.
10. Ibid., 53.
11. Serres and Latour, *Conversations*, 58.
12. Shields, "Functionalism."
13. Hausman, "Ovaries," 166–67.
14. Patrick, "Psychology," 4.
15. Ibid.
16. Mall, "On Several," 7.
17. Ibid., 24.
18. Ibid., 27.
19. Ibid., 32.
20. Patrick, "Psychology," 6.
21. Ibid., 9.
22. Möbius, "Physiological," 625–28.
23. Ibid., 629.
24. Ibid., 632–35.
25. Ibid., 636.
26. Shields, "Functionalism," 750.
27. Ibid., 751.
28. Dunlap, "Are There," 307.
29. Hollingworth, "Variability."
30. Möbius, "Physiological," 629.
31. Shields, "Functionalism," 745.
32. Jensen, "Improving."
33. Borell, "Organotherapy."
34. Ibid., 29.
35. Jensen, "Improving," 335.
36. Corner, "Our Knowledge."
37. Ibid., 919–20.
38. Ibid.
39. Marshall and Runciman, "Ovarian," 2.
40. Ibid., 21.
41. Allen, "Studies," 295.

42. Corner and Allen, "Physiology," 335.
43. Allen and Doisy, "Ovarian."
44. Corner and Allen, "Physiology," 338.
45. Frank, *Female*, foreword.
46. Stone, "Experimental," 102.
47. Young, "Sen. Bill Kintner."
48. "Forgiveness."

Chapter 6

1. Butenandt and Westphal, "Isolation," 140.
2. Micale, "On the 'Disappearance.'"
3. Serres and Latour, *Conversations*, 64.
4. Ibid., 66.
5. Gronnvoll and Landau, "From Viruses."
6. Koerber, "From Folklore."
7. Jensen, "From Barren."
8. Lakoff and Johnson, *Metaphors*, 6.
9. Keller, *Making*, 118–20.
10. Segal, *Health*, 118.
11. Jensen, "From Barren," 26.
12. Keller, *Making*, 120.
13. *Aristotle's Masterpiece*, 94.
14. Veith, *Hysteria*, 164–73.
15. Ibid., 196–97.
16. Ceccarelli, "Neither"; Jensen, "From Barren."
17. Keller, *Making*, 114.
18. Allen, "Studies," 295–97.
19. See Frank, *Female*, 198–11.
20. Frank, "Hormonal," 1053–57.
21. Frank, "Hormonal," 1053.
22. Ibid., 1054.
23. Frank, *Female*, 3.
24. Frank, "Hormonal," 1054.
25. Ibid., 1054–56.
26. Ibid., 1055.
27. Ibid., 1056.
28. Ibid.
29. Ibid., 1057.
30. Ibid.
31. Ibid.
32. Ceccarelli, "Neither"; Condit et al., "Recipes."
33. Whitehead, "Notes," 47.
34. Ibid.
35. Ibid., 48–49.
36. Seward, "Female," 172.
37. Ibid., 175–79.
38. Ibid., 179–80.

39. Ibid., 180.

40. Ibid., 182.

41. Benedek and Rubinstein, "Correlations: I," 245.

42. Ibid., 247.

43. Ibid., 249.

44. Ibid., 253.

45. Ibid., 259.

46. Ibid.

47. Ibid., 270.

48. Benedek and Rubinstein, "Correlations: II," 461.

49. Martin, *Flexible*.

50. Benedek and Rubinstein, "Correlations: II," 462–64.

51. Ibid., 485–86.

52. Greene, "Discussion," 337–38.

53. Ibid., 345.

54. Ibid., 339.

55. Ibid.

56. Ibid., 341.

57. Ibid., 343.

58. Ibid., 346–47.

59. Baake, *Metaphor*, 62.

60. Jensen, "From Barren."

61. Ceccarelli, "Neither."

62. Serres and Latour, *Conversations*, 65.

63. Keller, *Making*, 114.

Chapter 7

1. *Pregly*.

2. Aristotle's *Rhetoric*, book 1, chapter 2.

3. Burnyeat, "Enthymeme," 100.

4. Walker, "Body," 52–53.

5. Jackson, "Enthymematic," 616.

6. Hurt, "Legitimizing," 381–82.

7. Brett and Baxendale, "Motherhood," 342.

8. Ibid., 344.

9. Ibid., 346.

10. Ibid., 348.

11. Ibid., 351.

12. Ibid., 354–55.

13. Ibid., 355.

14. Ibid., 357.

15. Henry and Rendell, "Review," 793.

16. Ibid., 794.

17. Ibid.

18. Ibid., 799.

19. Ibid., 800.

20. Hurt, "Legitimizing," 383.

21. Ibid., 392.
22. Ibid., 385–86.
23. Scott, "Public," 58.
24. Ibid., 61.
25. Moore, "Pregnant," 5.
26. Oatridge et al., "Change," 26.
27. Brett and Baxendale, "Motherhood," 340.
28. Genevieve, "'Baby Brain.'"
29. Edbauer, "Unframing."
30. Albert, Pruessner, and Newhouse, "Estradiol," 16–17.
31. Ibid., 16.
32. Ibid., 14.
33. Ibid., 22.
34. Walker, "Body," 61.
35. Ussher, "Are We Medicalizing," 13–14.
36. Derntl et al., "Significant."
37. Albert, Pruessner, and Newhouse, "Estradiol," 14.
38. Ussher, "Are We Medicalizing," 12.
39. Albert, Pruessner, and Newhouse, "Estradiol," 15.
40. Donner and Lowry, "Sex Differences," 2.
41. Olff et al., "Gender," 194.
42. Kinsley and Lambert, "Reproduction-Induced," 515–16.
43. Ibid., 522–23.
44. Glynn, "Increasing," 1038.
45. Ibid., 1042.
46. Ibid., 1043.
47. Thornton, "Transformations," 272.
48. Ibid., 273.
49. Ibid., 277.
50. Ibid., 278.
51. Ibid., 282–83.
52. Brett and Baxendale, "Motherhood," 358.
53. Henry and Rendell, "Review," 799.
54. Jackson, "Enthymematic," 606.
55. Walker, "Body," 48–49.
56. Serres and Latour, Conversations, 82.

Chapter 8

1. Miller, "Kairos."
2. Segal, Health.
3. See, for example, Reeves, "Rhetoric"; Segal, "Illness"; and Lynch, What.
4. Miller, "Kairos," 323.
5. Ibid., 311.
6. Serres and Latour, Conversations, 60–61.
7. Miller, "Kairos," 313.
8. Serres and Latour, Conversations, 58.
9. Kuhn, Structure; Feyerabend, "Attempt."

10. Campbell, "Anxiety"; Campbell, "Scientific."
11. Campbell, "Anxiety," 385.
12. Ballif, "Writing," 252.
13. Veith, *Hysteria*, viii.
14. Kinnane, "Metaphor."
15. Micale, "On the 'Disappearance,'" 499.
16. Ibid., 500–2.
17. Ibid., 511.
18. Ibid., 515.
19. Oudshoorn, *Beyond*, 82–111.
20. Roberts, *Messengers*, 44.
21. Ibid., 38.
22. Pinto, "Minding," 318.
23. E. Wilson, *Gut*, 28.
24. Alter, "Todd Akin."
25. Yan, "Donald."
26. Flegenheimer and Haberman, "Donald."
27. Culp-Ressler, "2013"; Arduser and Koerber, "Splitting."
28. Liptak, "Supreme."
29. Weber, "Texas."
30. Webb et al., "Developmental."
31. Morris, Song, and Hasemyer, "Fracking."
32. Koerber, Booher, and Rickly, "Concept."
33. Buzzanell, "Good."
34. Dubriwny, *Vulnerable*.
35. Buzzanell, "Good," 270.
36. Ibid., 274.
37. Thornton, "Neuroscience," 400–4.
38. Ibid., 408–9.
39. Ibid., 412.
40. Ussher, "Are We Medicalizing," 19.
41. Ibid.
42. Ibid., 25.
43. Ussher, *Managing*, 163.
44. Roberts, *Messengers*, 21.
45. Irni, "Sex," 53.
46. Ibid., 43.
47. E. Wilson, *Gut*, 5.
48. Ibid., 3.
49. Holland, *Moody*, 2.
50. Baird, "How."
51. Pentland, *Social*, loc 200 (Kindle version).
52. Baird, "How."
53. Serres and Latour, *Conversations*, 154.
54. Newman, "There Is."

Bibliography

Albert, Kimberly, Jens Pruessner, and Paul Newhouse. "Estradiol Levels Modulate Brain Activity and Negative Responses to Psychosocial Stress across the Menstrual Cycle." *Psychoneuroendocrinology* 59 (2015): 14–24.

Allen, Chauncey N. "Studies in Sex Differences." *Psychological Bulletin* 24 (May 1927): 294–304.

Allen, Edgar, and Edward A. Doisy. "An Ovarian Hormone: Preliminary Report on Its Localization, Extraction and Partial Purification, and Action in Test Animals." *Journal of the American Medical Association*, September 18, 1923, 819–21.

Alter, Charlotte. "Todd Akin Still Doesn't Get What's Wrong with Saying 'Legitimate Rape.'" *Time*, July 17, 2014. http://time.com/3001785/todd-akin-legitimate-rape-msnbc-child-of-rape/.

American Psychiatric Association. *Diagnostic and Statistical Manual of Mental Disorders.* 4th ed. Washington, D.C.: American Psychiatric Association, 1994.

———. *Diagnostic and Statistical Manual of Mental Disorders.* 5th ed. Washington, D.C.: American Psychiatric Association, 2013.

"Animal Extracts as Therapeutic Agents." *British Medical Journal* (June 1893): 1279.

Arduser, Lora, and Amy Koerber. "Splitting Women, Producing Biocitizens, and Vilifying Obamacare in the 2012 Presidential Campaign." *Women's Studies in Communication* 37 (2014): 117–37.

Aristotle's Masterpiece: The Midwife's Guide, Illustrated. New York: Published for the Trade, 1846.

Baake, Ken. *Metaphor and Knowledge: The Challenges of Writing Science.* Albany: SUNY Press, 2003.

Baird, Julia. "How to Explain Mansplaining." *New York Times*, April 20, 2016. http://mobile.nytimes.com/2016/04/21/opinion/how-to-explain-mansplaining.html.

Ballif, Michelle. "Writing the Event: The Impossible Possibility for Historiography." *Rhetoric Society Quarterly* 44 (2014): 243–55.

Bayliss, Sir William Maddock. *Principles of General Physiology.* London: Longmans, Green, and Company, 1920.

Benedek, Therese, and Boris B. Rubinstein. "The Correlations between Ovarian Activity and Psychodynamic Processes: I. The Ovulative Phase." *Psychosomatic Medicine* 1 (1939): 245–70.

———. "The Correlations between Ovarian Activity and Psychodynamic Processes: II. The Menstrual Phase." *Psychosomatic Medicine* 1 (1939): 461–85.

Bogousslavsky, Julien. "Hysteria after Charcot: Back to the Future." In *Following Charcot: A Forgotten History of Neurology and Psychiatry.* Edited by Julien Bogousslavsky, 137–61. Basel: Karger, 2011.

Bonduelle, Michel, and Toby Gelfand. "Hysteria behind the Scenes: Jane Avril at the Salpêtrière." *Journal of the History of the Neurosciences* 8, no. 1 (1999): 35–42.

Borell, Merriley. "Organotherapy and the Emergence of Reproductive Endocrinology." *Journal of the History of Biology* 18 (Spring 1985): 1–30.

Boss, Jeffrey M. N. "The Seventeenth-Century Transformation of the Hysteric Affection, and Sydenham's Baconian Medicine." *Psychological Medicine* 9 (1979): 222.

Boylan, Michael. "The Galenic and Hippocratic Challenges to Aristotle's Conception Theories." *Journal of the History of Biology* 17 (1984): 83–112.

Brett, Matthew, and Sallie Baxendale. "Motherhood and Memory: A Review." *Psychoneuroendocrinology* 26 (2001): 339–62.

Brown, Steven D. "Michel Serres: Science, Translation, and the Logic of the Parasite." *Theory, Culture & Society* 19 (2002): 1–27.

Brown-Séquard, Charles-Edouard. "The Effects Produced on Man by Subcutaneous Injections of a Liquid Obtained from the Testicles of Animals." *The Lancet* (July 1889): 105–7.

Bullough, Vern L. "An Early American Sex Manual, or, Aristotle Who?" *Early American Literature* 7 (1973): 236–46.

Burnyeat, M. F. "Enthymeme: Aristotle on the Rationality of Rhetoric." In *Essays on Aristotle's Rhetoric*. Edited by Amélie Oksenberg Rorty, 152–202. Berkeley: University of California Press, 1996.

Butenandt, Adolf, and Ulrich Westphal. "Isolation of Progesterone—Forty Years Ago." *American Journal of Obstetrics and Gynecology* 120 (1974): 138–41.

Buzzanell, Patrice M. "The Good Working Mother: Managerial Women's Sensemaking and Feelings about Work-Family Issues." *Communication Studies* 56 (2005): 261–85.

Campbell, John Angus. "The 'Anxiety of Influence'—Hermeneutic Rhetoric and the Triumph of Darwin's Invention over Incommensurability." In *Rhetoric and Incommensurability*. Edited by Randy Allen Harris, 334–90. West Lafayette, Ind.: Parlor Press, 2005.

———. "Scientific Discovery and Rhetorical Invention: The Path to Darwin's *Origin*." In *The Rhetorical Turn: Invention and Persuasion in the Conduct of Inquiry*. Edited by Herbert W. Simons, 58–90. Chicago: University of Chicago Press, 1990.

Ceccarelli, Leah. "Neither Confusing Cacophony nor Culinary Complements: A Case Study of Mixed Metaphors for Genomic Science." *Written Communication* 21 (2004): 92–105.

———. *On the Frontier of Science: An American Rhetoric of Exploration and Exploitation*. East Lansing: Michigan State University Press, 2013.

Charcot, Jean-Martin. *Lectures on the Diseases of the Nervous System*. Translated by George Sigerson. London: New Sydenham Society, 1881.

Condit, Celeste M. *The Meanings of the Gene*. Madison: University of Wisconsin Press, 1999.

———. "'Mind the Gaps': Hidden Purposes and Missing Internationalism in Scholarship on the Rhetoric of Science and Technology in Public Discourse." *Poroi: An Interdisciplinary Journal of Rhetorical Analysis and Invention* 9, no. 1 (2013). http://dx.doi.org/10.13008/2151-2957.1150.

Condit, Celeste M., Benjamin R. Bates, Ryan Galloway, Sonja Brown Givens, Caroline K. Haynie, and John W. Jordan. "Recipes or Blueprints for Our Genes: How Contexts Selectively Activate the Multiple Meanings of Metaphors." *Quarterly Journal of Speech* 88 (2002): 303–25.

Connor, Steven. "Michel Serres's Milieux." Accessed October 15, 2015. http://www.stevenconnor.com/milieux/.

Corner, George W. "Our Knowledge of the Menstrual Cycle, 1910–1950." *The Lancet* (April 1951): 919–23.

Corner, George W., and Willard M. Allen. "Physiology of the Corpus Luteum." *American Journal of Physiology* 88 (March 1929): 326–39.

Culp-Ressler, Tara. "2013 Is Shaping Up to Be the Worst Year for Reproductive Freedom in Recent History." *ThinkProgress*. June 13, 2013. https://thinkprogress.org/2013-is-shaping-up-to-be-the-worst-year-for-reproductive-freedom-in-recent-history-9191b cd1b7fa#.2hzql2nw1.

DeBondt, T., Y. Jacquemyn, W. Van Hecke, J. Sijbers, S. Sunaert, and P. M. Parizel. "Regional Gray Matter Volume Differences and Sex Hormone Correlations as a Function of Menstrual Cycle Phase and Hormonal Contraceptives Use." *Brain Research* 1530 (2013): 22–31.

Demand, Nancy. *Birth, Death, and Motherhood in Classical Greece*. Baltimore: Johns Hopkins University Press, 1994.

Derkatch, Colleen. *Bounding Biomedicine: Evidence and Rhetoric in the New Science of Alternative Medicine*. Chicago: University of Chicago Press, 2016.

Derntl, Birgit, et al. "Significant Association of Testosterone and Prefrontal Stress Is Specific for Males." Poster presentation, conference of the Organization for Human Brain Mapping, Honolulu, June 2015. Accessed June 11, 2016. https://ww4.aievolution.com/hbm1501/index.cfm?do=abs.viewAbs&abs=3737.

Didi-Huberman, Georges. *Invention of Hysteria: Charcot and the Photographic Iconography of the Salpêtrière*. Translated by Alisa Hartz. Cambridge, Mass.: MIT Press, 2003. (Originally published 1982.)

Donner, Nina C., and Christopher A. Lowry. "Sex Differences in Anxiety and Emotional Behavior." *Pflugers Archiv: European Journal of Physiology* 465 (2013): 601–26.

Dubriwny, Tasha N. *The Vulnerable Empowered Woman: Feminism, Postfeminism, and Women's Health*. New Brunswick: Rutgers University Press, 2012.

Dunlap, Knight. "Are There Any Instincts?" Conference presentation, American Psychological Association, Cambridge, Mass., December 29, 1919.

Edbauer, Jenny. "Unframing Models of Public Distribution: From Rhetorical Situation to Rhetorical Ecologies." *Rhetoric Society Quarterly* 35 (2005): 5–24.

Emmons, Kimberly K. *Black Dogs and Blue Words: Depression and Gender in the Age of Self-Care*. New Brunswick: Rutgers University Press, 2010.

Eskin, Catherine R. "Hippocrates, *Kairos*, and Writing in the Sciences." In *Rhetoric and Kairos: Essays in History, Theory, and Practice*. Edited by Phillip Sipiora and James S. Baumlin. Albany: State University of New York Press, 2002.

Faber, Diana P. "Jean-Martin Charcot and the Epilepsy-Hysteria Relationship." *Journal of the History of the Neurosciences* 6 (1997): 275–90.

Feudtner, Chris. *Bittersweet: Diabetes, Insulin, and the Transformation of Illness*. Chapel Hill: University of North Carolina Press, 2003.

Feyerabend, Paul. "An Attempt at a Realistic Interpretation of Experience." *Proceedings of the Aristotelian Society* 32 (supplement): 75–104.

Fissell, Mary E. "Hairy Women and Naked Truths: Gender and the Politics of Knowledge in *Aristotle's Masterpiece*." *William and Mary Quarterly* 60 (January 2003): 43–74.

Flegenheimer, Matt, and Maggie Haberman. "Donald Trump, Abortion Foe, Eyes 'Punishment' for Women, then Recants." *New York Times*, March 30, 2016. http://www.nytimes.com/2016/03/31/us/politics/donald-trump-abortion.html.

"Forgiveness." YouTube video, 0:30, posted by Todd Akin, Republican senatorial candidate from Missouri, August 21, 2012. https://www.youtube.com/watch?v=R57E3S8RO7A.

Foucault, Michel. *The Birth of the Clinic: An Archeology of Medical Perception*. New York: Vintage Books, 1994.

Frank, Robert T. *The Female Sex Hormone*. Springfield, Ill.: Charles T. Thomas, 1929.

————. "The Hormonal Causes of Premenstrual Tension." *Archives of Neurology and Psychiatry* 26 (1931): 1053–57.

Genevieve. "'Baby Brain' Is Real—5 Crazy Ways Pregnancy Changes Your Brain." *Mama Natural.* February 1, 2016. http://www.mamanatural.com/baby-brain/.

Glynn, Laura M. "Increasing Parity Is Associated with Cumulative Effects on Memory." *Journal of Women's Health* 21 (2012): 1038–45.

Graham, S. Scott, and Carl G. Herndl. "Talking Off-Label: The Role of Stasis in Transforming the Discursive Formation of Pain Science." *Rhetoric Society Quarterly* 41 (2011): 145–67.

Greene, Raymond. "Discussion of the Premenstrual Syndrome." *Proceedings of the Royal Society of Medicine* 48 (1954): 337–47.

Gronnvoll, Marita, and Jamie Landau. "From Viruses to Russian Roulette to Dance: A Rhetorical Critique and Creation of Genetic Metaphors." *Rhetoric Society Quarterly* 40 (2010): 46–70.

Gross, Alan G., Joseph E. Harmon, and Michael Reidy. *Communicating Science: The Scientific Article from the 17th Century to the Present.* Oxford: Oxford University Press, 2002.

Happe, Kelly E. *The Material Gene: Gender, Race, and Heredity after the Human Genome Project.* New York: New York University Press, 2013.

Harris, Randy Allen, ed. *Rhetoric and Incommensurability.* West Lafayette, Ind.: Parlor Press, 2005.

Harvey, William. *On the Motion of the Heart and Blood in Animals.* Frankfurt, 1628.

Hausman, Bernice L. "Ovaries to Estrogen: Sex Hormones and Chemical Femininity in the 20th Century." *Journal of Medical Humanities* 20 (1999): 165–76.

Henderson, John. "Ernest Starling and 'Hormones': An Historical Commentary." *Journal of Endocrinology* 184, no. 1 (2005): 5–10.

Henry, Julie D., and Peter G. Rendell. "A Review of the Impact of Pregnancy on Memory Function." *Journal of Clinical and Experimental Neuropsychology* 29 (2007): 793–803.

Holland, Julie. *Moody Bitches: The Truth about the Drugs You're Taking, the Sleep You're Missing, the Sex You're Not Having, and What's Really Making You Crazy.* New York: Penguin, 2015.

Hollick, Frederick. *The Diseases of Woman: Their Causes and Cure Familiarly Explained.* New York: Burgess, Stringer, 1847.

————. *The Origin of Life and Process of Reproduction in Plants and Animals, with the Anatomy and Physiology of the Human Generative System, Male and Female, and the Causes, Prevention and Cure of the Special Diseases to Which It Is Liable.* Philadelphia: David McKay, 1902.

Hollingworth, Leta Stetter. "Variability as Related to Sex Differences in Achievement: A Critique." *American Journal of Sociology* 19 (January 1914): 510–30.

Huntley, Katie, and Lisa McClain. "Legitimately Ancient Ideas about Rape: The Roots of Todd Akin's Medical Beliefs." *The Blue Review: Scholarship in the Public Interest.* October 12, 2012. https://thebluereview.org/ancient-akin-rape/.

Hurt, Nicole Emily. "Legitimizing 'Baby Brain': Tracing a Rhetoric of Significance through Science and the Mass Media." *Communication and Critical/Cultural Studies* 8 (2011): 376–98.

InfoSpace Holdings LLC. "10 Things Men Should Know about Female Hormones." HowStuffWorks. Accessed October 9, 2015. http://health.howstuffworks.com/sexual-health/female-reproductive-system/10-things-about-female-hormones.htm.

Irni, Sari. "Sex, Power and Ontology: Exploring the Performativity of Hormones." *NORA—Nordic Journal of Feminist and Gender Research* 21 (2013): 41–56.

Jackson, Matthew. "The Enthymematic Hegemony of Whiteness: The Enthymeme as Anti-racist Rhetorical Strategy." *JAC* 26 (2006): 601–41.

Jensen, Robin E. "From Barren to Sterile: The Evolution of a Mixed Metaphor." *Rhetoric Society Quarterly* 45 (2015): 25–46.

———. "Improving upon Nature: The Rhetorical Ecology of Chemical Language, Reproductive Endocrinology, and the Medicalization of Infertility." *Quarterly Journal of Speech* 101 (2015): 331.

Johnson, Walter M. B. *An Essay on Diseases of Young Women.* London: Simkin Marshall, 1849.

Keller, Evelyn Fox. *Making Sense of Life: Explaining Biological Development with Models, Metaphors, and Machines.* Boston: Harvard University Press, 2003.

Keränen, Lisa. *Scientific Characters: Rhetoric, Politics, and Trust in Breast Cancer Research.* Tuscaloosa: University of Alabama Press, 2010.

Kinnane, Garry. "Metaphor, Pathography, and Hysteria: Recent American Writing about Illness." *Critical Review* 40 (2000): 91–107.

Kinsley, Craig H., and Kelly G. Lambert. "Reproduction-Induced Neuroplasticity: Natural Behavioural and Neuronal Alterations Associated with the Production and Care of Offspring." *Journal of Neuroendocrinology* 20 (2008): 515–25.

Kliff, Sarah. "Rep. Todd Akin Is Wrong about Rape and Pregnancy, but He's Not Alone." *The Washington Post*, August 20, 2012. https://www.washingtonpost.com/blogs/ezra-klein/wp/2012/08/20/rep-todd-akin-is-wrong-about-rape-and-pregnancy-but-hes-not-alone/.

Koerber, Amy. "From Folklore to Fact: The Rhetorical History of Breastfeeding and Immunity, 1950–1997." *Journal of Medical Humanities* 27 (2006): 151–66.

Koerber, Amy, Amanda K. Booher, and Rebecca J. Rickly. "The Concept of Choice as Phallusy: A Few Reasons Why We Could Not Agree More." *Present Tense* 2, no. 2 (2012). http://www.presenttensejournal.org/volume-2/the-concept-of-choice-as-phallusy-a-few-reasons-why-we-could-not-agree-more/.

Kotchen, Theodore A. "Historical Trends and Milestones in Hypertension Research: A Model of the Process of Translational Research." *Hypertension* 58 (2011): 522–38.

Kuhn, Thomas S. *The Structure of Scientific Revolutions.* Chicago: University of Chicago Press, 1962.

Lakoff, George, and Mark Johnson. *Metaphors We Live By.* Chicago: University of Chicago Press, 2008.

Latour, Bruno. "The Enlightenment without the Critique: A Word on Michel Serres' Philosophy." In *Contemporary French Philosophy.* Edited by J. Phillips Griffiths, 83–97. Cambridge: Cambridge University Press, 1987.

Laurence, William L. "New 'Tree of Life' Found by Chemists." *New York Times*, September 14, 1933.

Lawrence, Leah. "Controversial 'Father' of Endocrinology: Brown-Séquard." *Endocrine Today* (February 2008): 1. http://www.healio.com/endocrinology/news/print/endocrine-today/%7Bob179ieo-oeic-42ac-bec3-d5b67bb054f6%7D/controversial-father-of-endocrinology-brown-squard.

Liptak, Adam. "Supreme Court Strikes Down Texas Abortion Restrictions." *New York Times*, June 27, 2016. http://www.nytimes.com/2016/06/28/us/supreme-court-texas-abortion.html.

Lynch, John. *What Are Stem Cells?: Definitions at the Intersection of Science and Politics.* Tuscaloosa: University of Alabama Press, 2011.

MacDonald, Michael. *Witchcraft and Hysteria in Elizabethan London.* London: Routledge, 1991.

Maines, Rachel P. *The Technology of Orgasm: "Hysteria," the Vibrator, and Women's Sexual Satisfaction*. Baltimore: Johns Hopkins University Press, 2001.

Mall, Franklin P. "On Several Anatomical Characters of the Human Brain, Said to Vary According to Race and Sex, with Especial Reference to the Weight of the Frontal Lobe." *American Journal of Anatomy* 9 (1909): 1–32.

Martin, Emily. *Flexible Bodies: Tracking Immunity in American Culture from the Days of Polio to the Days of AIDS*. Boston: Beacon Press, 1994.

———. *The Woman in the Body: A Cultural Analysis of Reproduction*. Boston: Beacon Press, 1987.

Marshall, F. H. A., and J. G. Runciman. "On the Ovarian Factor Concerned in the Recurrence of the Oestrus Cycle." *Journal of Physiology* 49 (December 22, 1914): 17–22.

Merskey, Harold, and Susan J. Merskey. "Hysteria, or the Suffocation of the Mother." *Canadian Medical Association Journal* 148 (1993): 399–405.

Micale, Mark S. "On the 'Disappearance' of Hysteria: A Study in the Clinical Deconstruction of a Diagnosis." *Isis* 84 (1993): 496–526.

Miller, Carolyn R. "The Aristotelian *Topos*: Hunting for Novelty." In *Rereading Aristotle's Rhetoric*. Edited by Alan G. Gross and Arthur E. Walzer, 130–46. Carbondale: Southern Illinois University Press, 2000.

———. "*Kairos* in the Rhetoric of Science." In *A Rhetoric of Doing: Essays on Written Discourse in Honor of James L. Kinneavy*. Edited by James L. Kinneavy, Stephen Paul Witte, Neil Nakadate, and Roger Dennis Cherry, 310–27. Carbondale: Southern Illinois University Press, 1992.

Möbius, Paul J. "The Physiological Mental Weakness of Woman." *Alienist and Neurologist: A Quarterly Journal of Scientific, Clinical and Forensic Psychiatry and Neurology* 22 (1901): 624–42.

Moore, Pete. "Pregnant Women Get That Shrinking Feeling." *New Scientist* 153 (January 1997): 5.

Morris, Jim, Lisa Song, and David Hasemyer. "Fracking the Eagle Ford Shale: Big Oil and Bad Air on the Texas Prairie." *The Weather Channel*. February 18, 2014. http://stories.weather.com/fracking.

Newman, Jonah. "There Is a Gender Pay Gap in Academe, but It May Not Be the Gap That Matters." *The Chronicle of Higher Education*. April 11, 2014. http://chronicle.com/blogs/data/2014/04/11/there-is-a-gender-pay-gap-in-academe-but-it-may-not-be-the-gap-that-matters/.

Ng, Beng-Yeong. "Hysteria: A Cross-Cultural Comparison of Its Origins and History." *History of Psychiatry* 10 (1999): 287–301.

Oatridge, Angela, Anita Holdcroft, Nadeem Saeed, Joseph V. Hajnal, Basant K. Puri, Luca Fusi, and Graeme M. Bydder. "Change in Brain Size during and after Pregnancy: Study in Healthy Women and Women with Preeclampsia." *American Journal of Neuroradiology* 23 (2002): 19–26.

Olff, Miranda, Willie Langeland, Nel Draijer, and Berthold P. R. Gersons. "Gender Differences in Posttraumatic Stress Disorder." *Psychological Bulletin* 133 (2007): 183–204.

Oudshoorn, Nelly. *Beyond the Natural Body: An Archeology of Sex Hormones*. London: Routledge, 1994.

Patrick, George Thomas White. "The Psychology of Woman." *Popular Science Monthly* 47 (June 1895): 1–13.

Pavlov, Ivan. "Nobel Lecture: Physiology of Digestion." Nobel lecture, December 12, 1904. Accessed February 10, 2016. http://www.nobelprize.org/nobel_prizes/medicine/laureates/1904/pavlov-lecture.html.

Pentland, Alex. *Social Physics: How Good Ideas Spread—The Lessons Learned from a New Science*. New York: Penguin, 2014.

Pettit, Michael. "Becoming Glandular: Endocrinology, Mass Culture, and Experimental Lives in the Interwar Age." *American Historical Review* (October 2013): 1052–76.

Pinto, Pedro. "Minding the Body, Sexing the Brain: Hormonal Truth and the Post-Feminist Hermeneutics of Adolescence." *Feminist Theory* 13 (2012): 305–23.

Pregly. Last modified September 2011. Accessed June 8, 2016, http://www.pregnancythis week.com/forum/discussion/54151/its-not-my-fault-i-have-pregnancy-brain/p1.

Prelli, Lawrence J. *A Rhetoric of Science: Inventing Scientific Discourse*. Columbia: University of South Carolina Press, 1989.

Reeves, Carol. "Rhetoric and the AIDS Virus Hunt." *Quarterly Journal of Speech* 84 (1998): 1–22.

Rickert, Thomas. *Ambient Rhetoric: The Attunements of Rhetorical Being*. Pittsburgh: University of Pittsburgh Press, 2013.

Roberts, Celia. *Messengers of Sex: Hormones, Biomedicine, and Feminism*. Cambridge: Cambridge University Press, 2007.

Rowntree, Leonard G. "An Evaluation of Therapy, with Special Reference to Organotherapy." Presidential address, Society for the Study of Internal Secretion, Atlantic City, N.J. *Endocrinology* 9 (May 1925): 181–191.

Sajous, Charles E. "Endocrinology as a Key to the Solution of Major Medical Problems." *American Journal of the Medical Sciences* 164 (November 1922): 625–45. Printed version of a lecture read at the first scientific session of the Endocrinological Society of the City of New York. Accessed February 17, 2016. https://play.google.com/books/reader?id=JvRGAQAAMAAJ&printsec=frontcover&output=reader&hl=en&pg=GBS.PA625.

Schiebinger, Londa. *Nature's Body: Gender in the Making of Modern Science*. Boston: Beacon Press, 1993.

Schiller, Francis. *A Möbius Strip: Fin-de-Siècle Neuropsychiatry and Paul Möbius*. Berkeley: University of California Press, 1982.

Scott, J. Blake. "The Public Policy Debate over Newborn HIV Testing: A Case Study of the Knowledge Enthymeme." *Rhetoric Society Quarterly* 32 (2002): 57–83.

Scull, Andrew. *Hysteria: The Disturbing History*. Oxford: Oxford University Press, 2009.

Segal, Judy Z. *Health and the Rhetoric of Medicine*. Carbondale: University of Illinois Press, 2005.

———. "Illness as Argumentation: A Prolegomenon to the Rhetorical Study of Contestable Complaints." *Health: An Interdisciplinary Journal for the Social Study of Health, Illness and Medicine* 11 (2007): 227–44.

Seigel, Marika. *The Rhetoric of Pregnancy*. Chicago: University of Chicago Press, 2013.

Serres, Michel. *A History of Scientific Thought: Elements of a History of Science*. Oxford, U.K.: Blackwell, 1989.

———. *Rome: The Book of Foundations*. Translated by Felicia McCarren. Stanford: Stanford University Press, 1991.

Serres, Michel, and Bruno Latour. *Conversations on Science, Culture, and Time*. Ann Arbor: University of Michigan Press, 1995.

Seward, Georgene H. "The Female Sex Rhythm." *Psychological Bulletin* 31 (1934): 153–92.

Shields, Stephanie. "Functionalism, Darwinism, and the Psychology of Women." *American Psychologist* (July 1975): 745.

Slater, Eliot. "Diagnosis of Hysteria." *British Medical Journal* (May 1965): 1395.

Starling, Ernest Henry. *Croonian Lectures on the Chemical Correlation of the Functions of the Body: Delivered before the Royal College of Physicians of London on June 20th, 22nd, 27th, and 29th, 1905.* London: Printed by the Women's Printing Society [1905]. Accessed February 8, 2016. https://ia601301.us.archive.org/31/items/b2497626x/b2497626x.pdf.

Stewart, Francis E. "The Growing Importance of Endocrinology and Organotherapy." Twenty-Sixth Annual Meeting of the Seaboard Medical Association, Norfolk, Va., December 6–8, 1921. Published in *American Medicine* (January 1922): 17–24.

Stone, Calvin P. "Experimental Studies of Two Important Factors Underlying Masculine Sexual Behavior: The Nervous System and the Internal Secretion of the Testis." *Journal of Experimental Psychology* 6 (April 1923): 85–106.

Stormer, Nathan. *Sign of Pathology: US Medical Rhetoric on Abortion, 1800s–1960s.* University Park: Penn State University Press, 2015.

Tasca, Cecilia, Mariangela Rapetti, Mauro Giovanni Carta, and Bianca Fadda. "Women and Hysteria in the History of Mental Health." *Clinical Practice and Epidemiology in Mental Health* 8 (2012): 110–19. http://www.ncbi.nlm.nih.gov/pmc/articles/PMC3480686/.

Teston, Christa B. "A Grounded Investigation of Genred Guidelines in Cancer Care Deliberations." *Written Communication* 26 (2009): 320–48.

Teston, Christa, and S. Scott Graham. "Stasis Theory and Meaningful Public Participation in Pharmaceutical Policy." *Present Tense* 2 (2012). http://www.presenttensejournal.org/wp-content/uploads/2012/10/Teston.pdf.

Thompson, Lana. *The Wandering Womb: A Cultural History of Outrageous Beliefs about Women.* Amherst, N.Y.: Prometheus Books, 1999.

Thornton, Davi. "Transformations of the Ideal Mother: The Story of Mommy Economicus and Her Amazing Brain." *Women's Studies in Communication* 37 (2014): 271–91.

Thornton, Davi Johnson. "Neuroscience, Affect, and the Entrepreneurialization of Motherhood." *Communication and Critical Cultural Studies* 8 (2011): 399–424.

"Topology." Wolfram MathWorld. Accessed October 9, 2015. http://mathworld.wolfram.com/Topology.html.

Ussher, Jane M. "Are We Medicalizing Women's Misery? A Critical Review of Women's Higher Rates of Reported Depression." *Feminism and Psychology* 20 (2010): 9–35.

———. *Managing the Monstrous Feminine: Regulating the Reproductive Body.* London: Routledge, 2006.

Veith, Ilza. *Hysteria: The History of a Disease.* Chicago: University of Chicago Press, 1965.

Vesalius, Andreas. *On the Workings of the Human Body.* Padua: School of Medicine, 1543.

Walker, Jeffrey. "The Body of Persuasion: A Theory of the Enthymeme." *College English* 56, no. 1 (1994): 46–65.

Webb, Ellen, Sheila Bushkin-Bedient, Amanda Cheng, Christopher D. Kassotis, Victoria Balise, and Susan C. Nagel. "Developmental and Reproductive Effects of Chemicals Associated with Unconventional Oil and Natural Gas Operations." *Reviews on Environmental Health* 29 (2014): 307–18.

Weber, Paul J. "Texas May Not Restore Lost Abortion Clinics Despite Ruling." *Lubbock Avalanche-Journal,* June 29, 2016. http://lubbockonline.com/texas/2016–06–29/texas-may-not-restore-lost-abortion-clinics-despite-ruling#.V6zCo_krLrd.

WebMD LLC. "Escape from Hormone Horrors—What You Can Do." WebMD, accessed October 10, 2015. http://www.webmd.com/women/features/escape-hormone-horrors-what-you-can-do.

Whitehead, R. E. "Notes from the Department of Commerce: Women *Pilots*." *Journal of Aviation Medicine* 5 (1934): 47–49.

Wilson, Elizabeth A. *Gut Feminism.* Durham: Duke University Press, 2015.

Wilson, Jean D. "Charles-Edouard Brown-Séquard and the Centennial of Endocrinology." *Journal of Clinical Endocrinology and Metabolism* 71 (1990): 1403–9.

Yan, Holly. "Donald Trump's 'Blood' Comment about Megyn Kelly Draws Outrage." *CNN Politics*. August 8, 2015. http://www.cnn.com/2015/08/08/politics/donald-trump-cnn-megyn-kelly-comment/.

Young, JoAnne. "Sen. Bill Kintner: In His Own Words." *Lincoln Journal Star*. May 28, 2013. http://journalstar.com/legislature/sen-bill-kintner-in-his-own-words/article_1a3dbc72-1933-5401-be3f-c8c6a8e10bd8.html?comment_form=true.

Zimmer, Carl. *Soul Made Flesh: The Discovery of the Brain—and How It Changed the World*. New York: Free Press, 2004.

Index